CONSTRUCTION QUALITY MANAGEMENT

Quality management is essential for facilitating the competitiveness of modern day commercial organisations. Excellence in quality management is a requisite for construction organisations who seek to remain competitive and successful. The challenges presented by competitive construction markets and large projects that are dynamic and complex necessitate the adoption and application of quality management approaches.

This new edition of *Construction Quality Management* provides a comprehensive evaluation of quality management systems and tools. Their effectiveness in achieving project objectives is explored, as well as applications in corporate performance enhancement. Both the strategic and operational dimensions of quality assurance are addressed by focusing on providing models of best practice.

The reader is supported throughout by concise and clear explanations and with self-assessment questions. Practical case study examples show how various evaluative-based quality management systems and tools have been applied. Subjects covered include:

- business objectives – the stakeholder satisfaction methodology
- organisational culture and Health and Safety
- quality philosophy
- evaluation of organisational performance
- continuous quality improvement and development of a learning organisation.

New chapters consider the influence of Building Information Modelling (BIM) on quality management. The text should be of interest to construction industry senior managers, practicing professionals and academics. It is also an essential resource for undergraduate and postgraduate students of construction management, project management and business management courses.

Tim Howarth is a learning and development specialist for a large international construction contractor. He has 20 years of experience as a construction management academic.

David Greenwood is Professor of Construction Management at Northumbria University, Newcastle upon Tyne, UK. He is also a Director of BIM Academy.

CONSTRUCTION QUALITY MANAGEMENT

Principles and Practice

Second edition

Tim Howarth and David Greenwood

Routledge
Taylor & Francis Group

LONDON AND NEW YORK

Second edition published 2018
by Routledge
2 Park Square, Milton Park, Abingdon, Oxon, OX14 4RN

and by Routledge
711 Third Avenue, New York, NY 10017

Routledge is an imprint of the Taylor & Francis Group, an informa business

First edition published by Spon Press 2011

Library of Congress Cataloging-in-Publication Data
Names: Howarth, Tim, author. | Greenwood, David, 1951 December
    27– author.
Title: Construction quality management : principles and practice / by
    Tim Howarth and David Greenwood.
Description: New York, NY : Routledge, 2018. | Revised edition of:
    Construction quality management : principles and practice / Paul Watson,
    Tim Howarth. 2011. | Includes bibliographical references and index.
Identifiers: LCCN 2017016235 | ISBN 9781138680104 (hardback : alk. paper) |
    ISBN 9781138680111 (pbk. : alk. paper) | ISBN 9781315563657 (ebook)
Subjects: LCSH: Building—Quality control.
Classification: LCC TH438.2 .H69 2018 | DDC 690.068/5—dc23
LC record available at https://lccn.loc.gov/2017016235

ISBN: 978-1-138-68010-4 (hbk)
ISBN: 978-1-138-68011-1 (pbk)
ISBN: 978-1-315-56365-7 (ebk)

Typeset in Bembo
by Apex CoVantage, LLC

CV 08.15.2019 1245

CONTENTS

FIGURES

TABLES

ABBREVIATIONS

BIM	Building Information Modelling
CBPP	Construction Best Practice Programme
CDM	Construction Design and Management Regulations
CIB	Construction Industry Board
CMPS	Centre for Management and Policy Studies
EFQM	European Foundation for Quality Management
HSE	Health and Safe Executive
ILO	International Labour Office
IOSH	Institution of Occupational Safety and Health
ISO	International Organisation for Standards
KPI	Key Performance Indicator
MFAM	Management Functional Assessment Model
OHSAS	Occupational Health and Safety Assessment Series
OHSMS	Occupational Health and Safety Management Systems
QA	Quality Assurance
QMS	Quality Management System
RADAR	Results Approach Deployment Assessment and Review
SEC	Specialist Engineering Contractors
SMEs	Small and Medium Sizes Enterprises
SMT	Senior Management Team
TQM	Total Quality Management

INTRODUCTION

This second edition presents a revised and updated review and discussion of quality management principles and practices in construction. The emergence of BIM has resulted in the inclusion of two new chapters: 'BIM as a quality management tool' and 'Assessing and demonstrating BIM capability'.

Learning outcomes are specified at the start of each chapter in order to enable the reader to readily identify the key principles addressed in each chapter. Each chapter includes a reading list so as to enable the reader to further explore and investigate topics of particular interest. Self-assessment questions at the end of each chapter support the reader in reflecting upon material and consolidating under-standing of the topics presented and discussed herein.

This book should prove useful to professionals and senior management in the construction industry as well as to university academics and students studying undergraduate and postgraduate construction management, project management and business management awards.

Chapter 1 An overview of the key theorists and quality philosophy

This chapter provides a succinct overview of key theories and people that have contributed significantly to the development of the concept and practice of quality management in modern-day commercial organisations.

Definitions and notions of quality are presented and the development of quality management is briefly outlined. The contributions of key proponents, theorists and pioneers of quality management are concisely outlined. Finally, the principles and philosophy of Total Quality Management (TQM) are explored and the advantages and problematic issues associated with implementing TQM within a commercial context are identified.

Chapter 2   Measuring project and corporate performance

This chapter explores advantages that are gained at both project and corporate levels by construction organisations fully engaging in the measurement of their performance. Improving project and corporate performance requires both the measurement of performance and the taking of actions that are informed by the feed forward of performance measurements.

Various methods of conducting performance measurement activities are presented and explained. Key measurement activities are also linked to the critical issue of obtaining stakeholder satisfaction at project and corporate level.

Chapter 3   Quality assurance and construction organisations

This chapter's focus is upon the philosophy and concept of quality assurance. Various critical aspects of the quality assurance are presented and explained for the reader.

Advocated advantages for construction organisations seeking quality assurance certification are established. In implementing a quality assurance system, construction organisations can encounter various problematic issues; these are identified and discussed. Suggestions are made as to how to avoid or address problematic issues that might be encountered.

Chapter 4   The European Foundation for Quality Management Excellence Model

This chapter provides an introduction to the philosophy, application and advocated advantages of TQM. Linkages between TQM and the European Foundation for Quality Management Excellence Model (EFQM.E.M) are explored and the constituent parts of the EFQM.E.M are outlined. Benefits of deployment of the excellence model within a construction context are outlined and issues associated with the application and deployment of the excellence model are identified and discussed.

Chapter 5   Developing organisational learning

This chapter considers project and corporate learning linked to continuous improvement. The chapter proposes that in order for construction organisations to fully engage in a continuous improvement process and strive for competitive advantage, they must develop the culture of a learning organisation. It is suggested that the concept of organisational learning be linked to the key functions of management; functions that serve to control organisation resources, procedures and systems. A self-assessment model that considers the management functions of construction organisations is outlined. This model serves to enable continuous improvement and excellence when linked with RADAR.

Chapter 6   Quality management systems for health and safety in construction

This chapter serves to inform of occupational health and safety management systems and outlines the essential components of such systems for organisations. Advocated benefits and problems associated with occupational health and safety management systems are indicated and differing standards and guidance documents are introduced. Examples of useful documentation for contributing to the systematic management and audit of health and safety on construction projects are provided at the end of the chapter.

Chapter 7   BIM as a quality system

This chapter treats BIM as a technology-enabled quality system. This system relates to the whole-life of a built asset and serves to improve the competitiveness of all organisations operating within the extended construction and property sectors. In considering BIM as a quality system, this chapter discusses the current state of development of BIM in the UK and bears in mind the fact that BIM is a rapidly moving global phenomenon.

Chapter 8   Assessing and demonstrating BIM capability

This chapter recognises that different organisations will adopt and adapt BIM differently and at different rates of progress. How the BIM maturity of an organisation might be demonstrated to clients, stakeholders and the world at large is considered. This issue is addressed via examination of how the quality of an organisation's BIM processes can be demonstrated. Starting with an outline of what criteria must be met, and how they can be measured, consideration is given to a number of emerging BIM certifying schemes and what it is they purport to certify.

1

AN OVERVIEW OF KEY THEORISTS AND QUALITY PHILOSOPHY

Introduction

This chapter presents a concise introduction to key theories and people that have contributed significantly to the development of the concepts and practices of quality management in modern-day organisations.

Various definitions and notions of quality are presented and the development of quality management practice in modern-day organisations is briefly outlined. The contributions of key proponents, theorists and pioneers of quality management are concisely outlined. Finally, the principles and philosophy of Total Quality Management (TQM) are explored and the advantages and problematic issues associated with implementing TQM within a commercial context are identified.

Learning outcomes

Upon completion of this chapter the reader will be able to demonstrate an understanding of:

- Differing definitions, notions and classifications of 'quality'.
- The contribution of seven key theorists to the development of quality within organisations.
- Key quality theories that inform and underpin the development and implementation of quality management approaches in modern-day organisations.
- Total Quality Management (TQM) and the advocated advantages and problematic issues associated with implementing TQM within a modern-day commercial context.

Defining quality

'Quality' is a word that is regularly applied and expressed within a great variety of contexts. In this modern-day commercial society, it is common to see advertisements that hold claims such as: 'premium quality', 'purveyors of quality', 'where quality comes first', 'superior quality', 'only the best quality materials', 'the place where quality counts' and so on.

It is difficult to contest that an association with the term 'quality' offers anything other than positive connotations. To be readily associated and affiliated with 'quality' and the notion of 'quality' is an aspiration of many modern-day organisations. Whilst being closely associated with 'quality' is entirely desirable to commercial organisations, establishing just what 'quality' means and what the quest to 'achieve quality' entails can be a matter open to some debate.

In a search for a definition of 'quality', Reeves and Bednar (1994) point out that

> the definition of quality has yielded inconsistent results. . . . [R]egardless of the time period or context in which quality is examined, the concept has had multiple and often muddled definitions and has been used to describe a wide variety of phenomena. Continued inquiry and research about quality and quality related issues must be built upon a thorough understanding of differing definitions of the construct.

When considering 'quality' as a term or concept it soon becomes apparent that it means many different things to many different people. There is quite clearly no one singular, universally accepted definition of 'quality'. The idea or concept of 'quality' is one that is multi-faceted. A survey of the 'definitions of quality' highlights this and is presented in *Table 1.1*. This survey identifies a range of suggested definitions and alternatives that serve to assist understanding, use and articulation of the term 'quality' within public and private sector organisations.

It is easy to identify from *Table 1.1* that there is no one singular, universally accepted definition of 'quality'. Indeed attempts to research and define quality within the commercial and organisational contexts of economics, manufacturing, the service industries and strategic and operations management have resulted in, as Garvin (1988) points out, a "host of competing perspectives each based on a different analytical framework, and employing its own terminology".

Whilst it can be recognised that definitions of quality are differing, they are not necessarily conflicting or contradictory. Rather, the diversity of definitions underlines the fact that quality is viewed in various ways. This diversity of views and definitions can be problematic, though – it can result in confused understanding, articulation and application of the quality concept within public and private sector organisations.

Classification of both the perspectives from which quality is viewed and the differing definitions of quality serves to clarify understanding regarding the quality concept. Such classification also serves to underpin and inform both communication and quality management practice. The following section offers an attempt at classifying quality definitions and serves to provide some meaning and structure to the diverse variety of quality definitions.

TABLE 1.1 Definitions of quality

Definition of quality – a thing is said to have the positive attribute of conformance to specified standards	Shewhart (1931)
Quality is a customer determination which is based on the customer's actual experience with the product or service, measured against his or her requirements – stated or unstated, conscious or merely sensed, technically operational or entirely subjective and always representing a moving target in a competitive market	Feigenbaum (1961)
Conformance to requirements	Crosby (1979)
Quality is (1) product performance which results in customer satisfaction (2) freedom from product deficiencies, which avoids customer dissatisfaction	Juran (1988)
Quality: the totality of features and characteristics of a product or service that bears on its ability to meet a stated or implied need	ISO 8402–1986, "Quality Vocabulary"
Quality is anything which can be improved	Masaaki (1986)
Quality is the loss a product causes to society after being shipped	Taguchi (1986)
Quality is the total composite product and service characteristics of marketing, engineering, manufacture and maintenance through which the product in use will meet the expectations of the customer	Feigenbaum (1951)
Good quality means a predictable degree of uniformity and dependability at a low cost with a quality suited to the market	Deming (1986)
Fitness for use	Juran (1988)
Quality is the extent to which the customer or users believe the product or service surpasses their needs and expectations	Gitlow et al. (1989)

Classifying the ways of looking at quality

The quality of a product or service can be viewed in purely objective or subjective terms, or in a manner that utilises both objective and subjective evaluation together. *Table 1.2* illustrates the classification of objective and subjective ways of viewing quality.

A research study undertaken by David Garvin (1986) drew upon surveys of 'first-line supervisors' in the USA and Japan and compared practices and attitudes concerning quality. Within this study Garvin identifies five distinct classifications for quality definitions. These five classifications are identified and expanded within *Table 1.3*.

Further to these classifications *Figure 1.1* illustrates Zhang's (2001) 'map of quality perspectives'. This brings together Garvin's five classifications of quality definitions and the objective and subjective measurement of quality. In *Figure 1.1* Zhang considers each of Garvin's five quality definition classifications in terms of:

- the extent of the *objective-subjective* determination of each classification of quality definition; and
- the *location* of *where* each quality definition classification is determined (internal or external to and organisation).

TABLE 1.2 Objective and subjective classifications of quality

Objective quality	Subjective quality
Here the concept of quality is grounded within the precept that the characteristics of a product or service are tangibly measurable and assessable in *absolute* terms such as size, design conformance, durability and performance.	Here the concept of quality is grounded in the *perceived* ability of a product or service to satisfy various needs and aspirations. Here each individual's perceptions can vary regarding the very same product or service.

TABLE 1.3 Garvin's five classifications of quality definitions

1	*Transcendental definition of quality*	Quality is viewed from a perspective of 'abstract properties', evaluated with innate knowledge gained from experience. In other words, "I can tell quality when I see it". Within this context, the determination of quality is *subjective* and is based upon 'the view of an individual', this view being developed with experience.
2	*Product-based definition of quality*	Quality is viewed from a perspective of 'desired attributes'. In this context, the prescribed *features* of a product, including its *performance*, serve to define its quality.
3	*User-based definition of quality*	Quality is viewed from a perspective of 'client/customer satisfaction'. In other words, quality relates to the extent to which client/customer needs and wants are satisfied by the 'fitness for purpose' of the service or product.
4	*Manufacturing-based definition of quality*	Quality is viewed from a perspective of 'manufacturing compliance'. In other words, 'a product's conformance to specified requirements'. Products are manufactured *within a tolerable scope* of variance. Where products are manufactured outside the tolerable scope of variance, they are scrapped or re-worked.
5	*Value-based definition of quality*	Quality is viewed from a perspective of 'economic utility'. In other words, is the service or product value for money? The determination of whether *value* is achieved is the subjective judgement of the client/customer. This 'value-based' classification of quality provides the grounding for service-sector research into the 'quality gap'.

Adapted from Garvin, 1986.

The perspectives from which quality may be viewed can be further classified in accordance with an organisation's *product* or *service* function. Product quality and service quality are commonly determined via consideration of differing criteria. Examples of these differing quality criteria are presented in *Tables 1.4* and *1.5*. In *Table 1.4* Garvin (1988) identifies and classifies eight dimensions of 'product quality'. This can be contrasted with *Table 1.5* where Parasuraman et al. (1988) identify and classify five dimensions of 'service quality' in their 'Servqual' model.

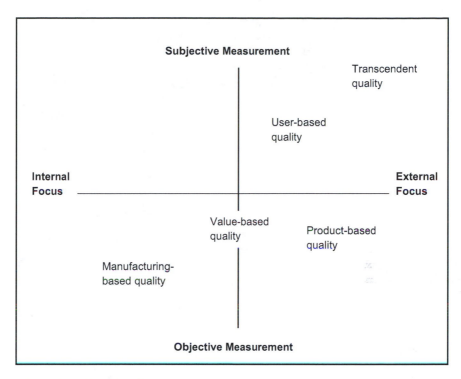

FIGURE 1.1 Map of quality perspectives

TABLE 1.4 Dimensions of product quality

Aesthetics	A subjective judgement of a product's look, feel, sound, taste, or smell.
Conformance	Compliance of a product's characteristics with predetermined physical and performance characteristics/standards.
Durability	Amount of use of a product before deterioration means that replacement is preferable or necessary. This can also be referred to as 'expected life'.
Features	The distinct properties of a product.
Perceived Quality	Subjective assessment of product quality. This is influenced by the product's brand name, image and associated advertising.
Performance	Product's primary characteristic of concern when considering the use or operation of the product. For example, 'miles per gallon' may be one such characteristic when considering a car, the 'tog value' (or thermal resistance) might be one such characteristic when considering a duvet', picture and sound clarity may be a key characteristic when considering a television.
Reliability	Product's ability to deliver to required standards, under stated conditions of use, over a specified period of time.
Serviceability	The ease and speed of repair and maintenance.

Adapted from Garvin (1988).

TABLE 1.5 Dimensions of service quality

Assurance	Employee's knowledge, courtesy and ability to inspire trust and confidence.
Empathy	Employee's ability to provide customers with caring individualised attention.
Reliability	Ability to deliver the promised service dependably and accurately.
Responsiveness	Ability to deliver the service promptly.
Tangibles	These include the appearance of physical facilities, equipment and personnel.

Adapted from Parasuraman et al. (1988).

Clearly quality is not a singular concept that can be viewed from only one perspective. It has a range of possible definitions and can be seen from a variety of perspectives (subjective, objective, user-based, value-based) and within a variety of contexts (service provision, product manufacture). This range of definitions and ways of looking at and classifying quality has served to develop, evolve and inform quality management practice throughout the twentieth century.

The development of quality management practice

The twentieth century saw dramatic growth in production and service industries and the realisation of the global market place and international business organisations.

Post-production inspection predominated quality management practice in the pre–World War II era. Significant changes and developments in quality management theory and practice were seen after World War II. *Figure 1.2* highlights key developments in quality management practice in the twentieth century with the presentation of 'timeline'. *Table 1.6* meanwhile summarises the key attributes and identifying characteristics of the various 'quality movements' of the century.

Key quality theorists and pioneers

A number of 'pioneers' have contributed significantly to the shaping and growth of quality management theory and practice. Each of the following seven 'theorists', 'gurus' or 'pioneers' can be recognised as having distinctively added to an aspect of understanding, development or operation of quality within the management of organisations. The seven notable quality management pioneers are:

- W Edwards Deming
- Joseph Juran
- Kaoru Ishikawa
- Armand V Feigenbaum
- Genichi Taguchi
- Philip Crosby
- Masaaki Imai

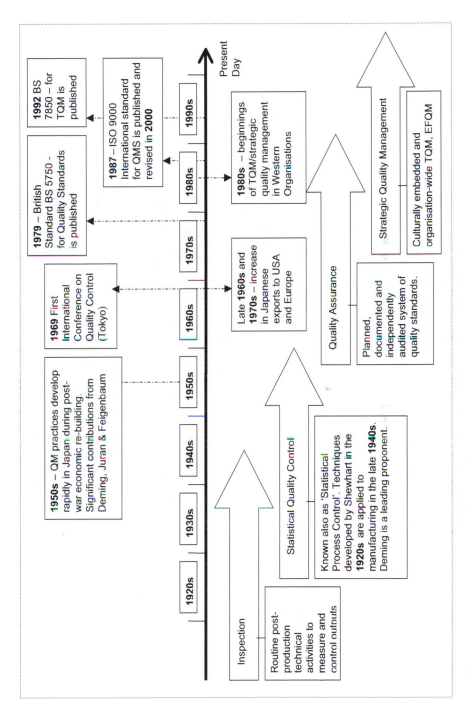

FIGURE 1.2 Timeline of the key developments in quality management practice

TABLE 1.6 Key attributes of quality management movements

Identifying Characteristics	Quality Movement			
	Inspection	Statistical Quality Control	Quality Assurance	Strategic Quality Management
Primary Concern	Detection	Control	Co-ordination	Strategic view
Emphasis	Product uniformity	Product uniformity with reduced inspection	The entire production system	The market and consumer needs
Methods	Measuring	Statistical tools and techniques	Procedural systems	Strategic planning and setting
Role of Quality Professionals	Inspection, Acceptance, Sampling	Trouble shooting and the application of statistical methods	Design of QA system, planning, measurement of performance audit	Goal setting, education, training and consultation
Responsibility for Quality	Inspection Department	Manufacturing, Engineering Departments	All departments	Everyone in the organisation
Orientation and Approach	Quality is 'inspected in'	Quality is 'controlled in'	Quality is 'built in'	Quality is 'managed in'

Adapted from Garvin (1988).

W Edwards Deming

William Edwards Deming was born in 1900 and between the years of 1917 and 1928 he enrolled on, and graduated from, a Bachelors degree in Electrical Engineering at the University of Wyoming, attained a Masters degree in Mathematics from the University of Colorado and gained a doctorate in mathematical physics from Yale University. Further to this Deming took a job at the United States Department of Agriculture where he was responsible for courses in mathematics and statistics.

In 1938 Deming took a position as an advisor in statistical sampling with the United States Government Service's Bureau of Census. Here he applied statistical methods to clerical-operations to establish sampling techniques for the 1940 census. Deming's work realised great improvements in productivity. As a result of his success he was retained in 1942 as a consultant by the War Department. He was later sent to Japan in 1946 by the War Department's Economic and Scientific section to study agricultural production. Here he made connection with the Union of Japanese Scientist and Engineers and was invited to deliver courses in statistical methods to Japanese industry. As a result, Deming delivered lectures in Japan throughout the 1950s on 'statistical methods' as a means of inculcating quality into industry output.

The Western world's recognition of Deming's contribution to quality through his work in Japan did not really come about until the 1980s. In this decade he published *Quality, Productivity, and Competitive Position* and *Out of Crisis*. Before his passing in 1993 Deming was the recipient of numerous awards and recognitions for his work. These awards include:

> The Second Order Medal of the Sacred Treasure – Japan's highest accolade to a foreign-national; The American Management Association's Taylor Award; and the National Medal of Technology – presented by President Ronald Reagan in 1987.

Deming's key concepts and contributions to quality theory

Deming's approach to quality is one that strives to 'delight customers'. It is an approach that is concisely portrayed in *Figure 1.3*.

The key concepts and contributions of Deming concern:

- The Plan, Do, Check, Act (PDCA) Cycle – a methodology for problem solving;
- Seven Deadly Diseases of Western Management;
- Fourteen Points for delivering transformation of an organisation for improved efficiency;
- A Seven Point Action Plan for change; and
- A System of Profound Knowledge.

FIGURE 1.3   Striving to delight customers

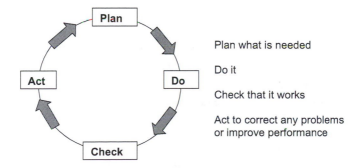

FIGURE 1.4   Deming's Plan, Do, Check, Act Cycle (PDCA Cycle)

Deming places a quality focus upon the causes of variation and variability in an organisation's manufacturing process. In striving to delight the customer, emphasis is put upon the production process, with the deployment of a statistical approach to measure the variability of a given process. For Deming variation is a key factor in poor quality and variation is the result of either a 'common cause' or a 'special cause'.

Common causes are defined as being systemic and arise from the design or operation of the production system. These causes of variation are viewed as being the responsibility of management. Special causes of variation on the other hand are evidenced at a local level by such things as the changing of an operator, shift or machine. These causes of variation are resolved by the giving of attention to each individual cause at the local level.

In Deming's view, management planning is essential if variation, wastage and selling price are to be reduced. The 'Plan, Do, Check, Act' Cycle underlines this necessity and provides a methodology for problem-solving and improvement.

Further to his work in Japan, promoting focus upon variance and the adoption of a systematic approach to problem solving – the PDCA Cycle – Deming identified problems or 'diseases' associated with organisations that required addressing. The 'Seven Deadly Diseases' are identified as:

- Lack of consistency of purpose.
- Emphasis on short term profit.
- Reliance on performance appraisal and merits.
- Staff mobility.
- Reliance on financial figures.
- Excessive medical costs.
- Excessive legal costs.

Deming's Fourteen Points, for management to enable efficiency within an organisation, were originally presented in his book *Out of Crises*. The Fourteen Points are applicable to all organisations and are:

1 Create constancy of purpose towards improvement of product and service, with the aim to become competitive and to stay in business, and to provide jobs.
2 Adopt the new philosophy. We can no longer live with commonly accepted levels of delay, mistakes and defective workmanship.
3 Cease dependence on inspection to achieve quality. Eliminate the need for mass inspection by building quality into the product.
4 End the practice of awarding business on the basis of price; instead minimise total cost and move towards single suppliers for items.
5 Improve constantly and forever the system of production and service to improve quality and productivity and to decrease costs
6 Institute training on the job.
7 Institute leadership; supervision should be to help to do a better job; overhaul supervision of management and production workers.
8 Drive out fear so that all may work effectively for the organisation.
9 Break down barriers between departments.

10 Eliminate slogans, exhortations and targets for the workforce asking for zero defects and new levels productivity.

    a   Eliminate work standards that prescribe quotas. Substitute leadership.

    b   Eliminate management by objective, by numbers and numerical goals. Substitute leadership.

11 Remove barriers that rob workers of their right to pride of workmanship (hourly workers; annual or merit rating, management by objective).

12 Remove barriers that rob people in management and engineering of their right to pride of workmanship.

13 Institute a vigorous education and self-improvement programme.

14 Put everybody in the organisation to work to accomplish the transformation. The transformation is everybody's job.

Deming provides organisations pursuing quality via the 'Fourteen Points' and contending with the 'Deadly Diseases' with a Seven Point Action Plan for change.

The steps of the Seven Point Action Plan are:

1 Management struggles over the 14 Points, Deadly Diseases and obstacles and agrees meaning and plans direction.

2 Management takes pride and develops courage for the new direction.

3 Management explains to the people in the company why change is necessary.

4 Divide every company activity into stages, identifying the customer of each stage as the next stage. Continual improvement of methods should take place at each stage, and stages should work together towards quality.

5 Start as soon and as quickly as possible to construct an organisation to guide continual quality improvement.

6 Everyone can take part in a team to improve the input and output of any stage.

7 Embark on construction of organisation for quality.

In his book *The New Economics*, Deming suggests that the prevailing management style of organisations requires a transformation and he proffers a 'System of Profound Knowledge' for this purpose. This system provides 'a map of theory by which to understand the organisations that we work in'. The System of Profound Knowledge is composed of four parts:

1 Appreciation for a system

2 Knowledge about variation

3 Theory of knowledge

4 Psychology.

Deming suggested that an individual holding an understanding of the System of Profound Knowledge would apply it to every kind of relationship with other people. Such a person would therefore:

- Set an example.
- Be a good listener, but will not compromise.

- Continually teach other people.
- Help people to pull away from their current practice and beliefs and move into the new philosophy without a feeling of guilt about the past.

Joseph Juran

Joseph Juran was born in 1904 in Braila in the east of Romania and was raised in Minnesota, USA. In 1924 he graduated with an Electrical Engineering degree from the University of Minnesota and began work as an engineer with the Western Electrical Company in Hawthorne, near Chicago.

In 1926 Western Electrical instituted a statistical quality control process and Juran became a founder employee of this department. He wrote *Statistical Methods Applied to Manufacturing Problems* in 1928, and by 1937 he was the Chief of Industrial Engineering at Western Electric's New York Office.

During the Second World War Juran served a four year leave of absence from Western. He was employed by the government as an administrator in the Lend-Lease Administration. He improved efficiencies within the department before leaving this Washington post in 1945. He did not return to Western Electrical, though, instead he began lecturing, writing and providing consultancy.

Juran joined New York University as a Head of Department and in 1951 he published *Quality Control Handbook*. In 1954 the Union of Japanese Scientists and Engineers invited him to lecture. He continued to lecture and contribute to Japan's economic development throughout the 50s and 60s and later published the lecturers he delivered in the 1964 book *Managerial Breakthrough*.

The 1970s brought the publication of 'Quality Planning and Analysis', and the founding of the Juran Institute – a training consultancy for the study of quality management – in 1979. Juran has since continued to actively publish his ideas and on 24 December 2004 he celebrated his 100th birthday.

Juran's key concepts and contributions to quality theory

For Juran quality is concerned with 'fitness for use or purpose', customers are located throughout the production process – both external and internal to an organisation, not simply at the end of the process – and 'breakthrough' to new levels of performance are required if an organisation is to survive and grow. Responsibility for quality is held by management and the awareness and training of management is essential to quality.

'Quality does not happen by accident, it must be planned' is a central view of Juran that is presented in 'On Planning for Quality'. Such planning is one component of his developed 'quality trilogy'. This trilogy consists of:

- Quality planning – designing a process that achieves required goals – this requires determining goals, undertaking resource planning, planning implementation and creating a quality plan;
- Quality control – operating and amend the process so as to achieve optimal effectiveness – monitoring performance, comparing achievements with

planned objectives and acting to close any gaps. Here a 'sensor' evaluates the performance of the system and reports to an 'umpire'. The umpire compares actual performance with the required goal and when significant discrepancy exists, the umpire reports to the 'actuator'. The actuator makes adjustments and changes to the system to ensure the achievement of the required goal; and

• Quality improvement – taking performance to new, superior levels – in terms of satisfying customers, reducing waste, enhancing logistics, improving employee morale and improving profitability.

The quality trilogy illustrated in *Figure 1.5*, places emphasis upon changing and developing the management of quality at the organisation's senior management level.

A 'Quality Planning Road Map' is provided by Juran for an organisation's measured implementation of each planning step. The roadmap details the following necessary steps:

• Identify who the customers are.
• Determine the needs of the identified customers.
• Translate those needs into the organisation's language.
• Develop a product that can respond to those needs.
• Optimise the product features so as to meet the organisation's needs as well as the needs of the customers.
• Develop a process which is able to produce the product.
• Optimise the process.
• Prove that the process can produce the product under operating conditions.
• Transfer the process to operations.

Further to this, Juran outlines a 'Formula for Results' – this states that an organisation must:

• Establish specific goals to be reached;
• Establish plans for reaching these goals;
• Assign clear responsibility for meeting the goals; and
• Base the rewards on results achieved.

FIGURE 1.5   Juran's quality trilogy

Kaoru Ishikawa

Kaoru Ishikawa was born in 1915. He studied at Tokyo University and graduated in 1939 with a Bachelor's degree in Applied Chemistry. In 1947 he became an assistant professor at the University.

Ishikawa was a founding member of the Japanese Union of Scientists and Engineers (JUSE) and attended lectures of Deming and Juran in Japan in the 1950s. By 1960 he had gained a doctorate and became a Professor. He contributed significantly to the development and implementation in the Japanese workplace of numerous tools of quality management, including:

- Quality Circles (the concept was initially published in the journal *Quality Control for the Foreman* in 1962);
- The Ishikawa Graph, also known as the Fishbone Graph, also known as the Cause and Effect Diagram; and
- The Seven Tools of Quality of Control – for use by workers within quality circles.

Ishikawa's contribution to quality within Japanese industry further extended with his roles of Chief Executive Director of the Quality Control Headquarters at JUSE and Chairman of the Editorial Committee of 'Quality Control for the Foreman'. He wrote and published two books before his passing in 1989 – *Guide to Quality Control* and *What Is Total Quality Control? The Japanese Way*.

Ishikawa's key concepts and contributions to quality theory

Ishikawa has developed the concept of 'company-wide quality control' and is widely regarded for his contribution to the Japanese 'quality circle movement' of the early 1960s and for the development of 'cause and effect', 'Ishikawa' diagrams.

With the deployment of 'quality circles', company-wide quality is advocated. The circles bring an inclusive, accessible and participative approach to quality within an organisation. This bottom-up approach to quality is varied in its application from organisation to organisation but generally consists of circles of between 4 and 12 worker-participants who identify local problems and recommend solutions.

The aims of quality circles are:

- to contribute to the improvement and development of the enterprise;
- to respect human relations and build a happy workshop; and
- to deploy human capabilities fully and draw out infinite potential.

Quality circles deploy seven statistical tools of quality control. These tools are taught to the organisation's employees and consist of:

- Pareto charts (to identify where the big problems are)
- Cause and effect diagrams (to identify what is causing the problems)
- Stratification (to show how the data is made up)

- Check sheets (to illustrate how often it occurs or is done)
- Histograms (to illustrate what variations look like)
- Scatter diagrams (to show relationships)
- Control charts (to identify which variations to control).

One of the tools deployed by the quality circles is the 'cause and effect' diagram, otherwise known as the 'fishbone' diagram or the 'Ishikawa' diagram. The Ishikawa diagram rather resembles a fishbone and is a systematic tool for investigating the causes of a particular effect and the relationships between cause and effect. The diagram is employed as a tool for identifying opinion regarding the most likely root cause for a specific, prescribed effect.

Ishikawa claims that when deploying company-wide quality control activities, the results are remarkable, in terms of ensuring the quality of industrial products contributing to the company's overall business.

Further to this, the effects of company-wide quality control are offered as:

- Product quality is improved and becomes uniform. Defects are reduced.
- Reliability of goods is improved.
- Cost is reduced.
- Quantity of production is increased, and it becomes possible to make rational production schedules.
- Wasteful work and rework are reduced.
- Technique is established and improved.
- Expenses for inspection and testing are reduced.
- Contracts between vendor and vendee are rationalised.
- The sales market is enlarged.
- Better relationships are established between departments.
- False data and reports are reduced.
- Discussions are carried out more freely and democratically.
- Meetings are operated more smoothly.
- Repairs and installation of equipment and facilities are done more rationally.
- Human relations are improved.

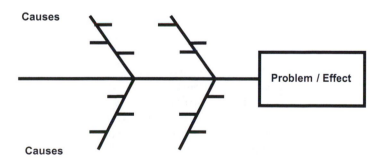

FIGURE 1.6 Ishikawa/Fishbone/cause and effect diagram

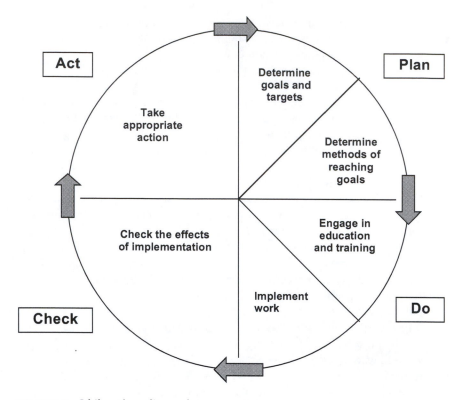

FIGURE 1.7 Ishikawa's quality cycle

In his 1985 book *What Is Total Quality Control?* Ishikawa expands Deming's PDCA Cycle of quality methodology from four steps into six. These steps are:

* Determine goals and targets.
* Determine methods of reaching goals.
* Engage in education and training.
* Implement work.
* Check the effects of implementation.
* Take appropriate action.

Armand V Feigenbaum

Armand Feigenbaum was born in 1920 and began his working life as an apprentice toolmaker at General Electric. He left General Electric to study for a BA in Industrial Administration at Union College, Schenectady, New York where he graduated in 1942 and attained a PhD at Massachusetts Institute of Technology in 1951. That same year he published his first book entitled *Quality Control: Principles and Practice*.

Quality Control: Principles and Practice was well received in Japan, at a time of industrial regeneration within the country. Feigenbaum's profile in Japan was further facilitated through his role as quality manager and later worldwide director of manufacturing operations with General Electric in the late 1950 and 1960s.

In 1961 Feigenbaum's second book was published, entitled *Total Quality Control*. This book reworked his earlier publication and is recognised as marking the first use of the term 'total quality control'. The year 1968 saw Feigenbaum leave General Electric to found General Systems Company – a quality management consultancy – with his brother Dr Donald Feigenbaum. Feigenbaum currently remains as President and Chief Executive Officer of General Systems Company, Inc located in Pittsfield, Massachusetts. Further to this, Feigenbaum was the founding chairman of the International Academy for Quality, has twice served as president of the American Society for Quality, was elected to the National Academy of Engineering of the United States in 1992 and has been the recipient of numerous awards and recognitions for his work with quality.

Feigenbaum's key concepts and contributions to quality theory

Feigenbaum is commonly considered as the originator of 'total quality control' – an approach to quality which advocates quality control as a comprehensive business method and demands quality-mindedness throughout an organisation. For Feigenbaum quality has now become an essential element of modern management – it is critical organisational success and company growth. His own definition of 'total quality control' is provided in his 1961 book of the same name:

> an effective system for integrating quality development, quality maintenance and quality improvement efforts of the various groups within an organisation, so as to enable production and service at the most economical levels that allow full customer satisfaction.

In attaining the business method of total quality control, three components are necessary:

- Quality leadership.
- Modern quality technology.
- Organisational commitment.

In this total quality context, 'control' is exercised throughout production as a management tool with four steps:

Step 1   Set quality standards.
Step 2   Appraise conformance to the standards.
Step 3   Act when standards are not attained.
Step 4   Plan to make improvements.

The 'Total Quality System' is defined by Feigenbaum as:

> The agreed company-wide and plant-wide operating work structure, documented in effective, integrated technical and managerial procedures, for guiding the co-ordinated actions of the people, the machines and the information of the company and plant in the best and most practical ways to assure customer quality satisfaction and economical costs of quality.

A management tool for measuring the total quality system is provided in the form of 'operating quality costs'. These are categorised as:

- Prevention costs – including quality planning.
- Appraisal costs – including inspection costs.
- Internal failure costs – including scrap and rework.
- External failure costs – including warranty costs and complaints.

More recently 'ten benchmarks for total quality success' have been defined by Feigenbaum. These benchmarks focus the organisation on the customer – both internal and external to the organisation. The benchmarks are:

1　Quality is a company-wide process.
2　Quality is what the customer says it is.
3　Quality and cost are a sum, not a difference.
4　Quality requires both individual and team zealotry.
5　Quality is a way of managing.
6　Quality and innovation are mutually dependent.
7　Quality is an ethic.
8　Quality requires continuous improvement.
9　Quality is the most cost-effective, least capital-intensive route to productivity.
10　Quality is implemented with a total system connected with customers and suppliers.

Genichi Taguchi

Genichi Taguchi was born in 1924. He was a student of textile engineering until he was drafted into the Astronomical Department of the Navigation Institute of the Imperial Japanese Navy between 1942 and 1945. After the war he was employed at the Ministry of Public Health and Welfare and the Institute of Statistical Mathematics.

In 1950 he took a position with the Nippon Telephone and Telegraph Company in the electrical communications laboratory. He became more widely known in this research and development role and served concurrently as a visiting professor at the Indian Statistical Institute between 1954 and 1955.

In 1962 Taguchi gained a Doctorate from Kyushu University, Japan and departed from the electrical communications laboratory of Nippon Telephone and Telegraph Company. He maintained consultancy links and became a Professor of Engineering at Aoyama Gakuin University in Tokyo in 1964, a position he held until 1982, when he became an advisor to the Japanese Standards Institute.

Taguchi served as Executive Director of the American Supplier Institute – a consulting organisation – and was an advisor to the Japanese Standards Institute.

Taguchi's key concepts and contributions to quality theory

Taguchi has developed an approach to quality that places emphasis upon the product design stage. The cornerstone of his approach is robust design achieved through methodical prototyped reduction of variance in the product. Put another way – Taguchi seeks reduction in variances accepted and tolerated in the production of a unit or item. To Taguchi, the tolerance of acceptable variance in a manufactured product results in what he describes as the 'loss function'.

For Taguchi, any variance from the exact product specification results in what he terms 'quality loss' – even through the product may still be within a traditionally accepted level of variance. This is illustrated in *Figure 1.8*.

Traditionally product components falling within the level of 'acceptable variance' would be considered useable. Taguchi views this as unacceptable in quality terms, as the failure of a product component to be exactly to target specification will facilitate customer dissatisfaction as the product may perform below its designed optimum. This could result in customer return of the product, customer refusal to purchase another product from the organisation and the customer advising others not to purchase the product. Taguchi clearly identifies the possibility of customer dissatisfaction with a product that is traditionally within tolerable levels of variance.

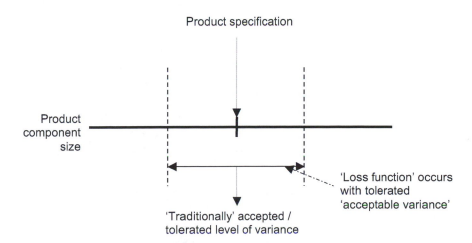

FIGURE 1.8   Taguchi's intolerance of variance – the loss function

Taguchi provides a quadratic formula for the 'loss function', this being:

$$L = c(x - T)^2 + k$$

Where:
X is a quality characteristic
T is the specified target
C is the cost of failing to meet the target
k is the minimum loss to society

FIGURE 1.9   Taguchi's loss function

The formula indicates that as product variation from the specified component target doubles, the quality loss is quadrupled. Dissatisfaction and loss is exponential to distance from the specified quality target.

Taguchi's approach to quality and reduction of the 'loss function' emphasises the need for optimisation of product and process *prior* to manufacture. He provides a methodology for the design of product tests prior to commencing manufacturing. As such, his approach is not one grounded in the pursuit of quality through inspection, instead his developed approach is one of employing 'off-line' quality control.

In pushing quality control back to the design stage Taguchi's methodology advocates the use of prototyping and experimental studies. An 'experimental design procedure' is prescribed for use within the design stage. This procedure provides for the efficient and effective carrying out of simulation experiments with the use of orthogonal arrays (OA) to enable the study of the simultaneous effect of several production process parameters and their interaction.

Reduction of quality loss is achieved via the deployment of a three stage prototyping methodology to the product manufacturing design process. The three stages are:

1   System design – for both the process and product.
2   Parameter design – investigate the optimum combination of process and product parameters. This is done with the objective of reducing sensitivity (and variance and loss factor) in production.
3   Tolerance design – identifies the sensitive components of the design that may be may give rise to variance in production. Alternatives are considered and tolerance limits are established.

By exercising quality control throughout the product manufacturing design stages, the Taguchi method aims to identify and reduce variance-causing 'noise' factors within the production process and 'optimise' production control factors.

Philip Crosby

Philip Crosby was born in West Virginia in 1926. He served in World War II, graduated from Western Reserve University and further serviced his nation in the Korean War. He began his civilian working life on the production line as a quality professional in 1952.

By 1965 Crosby had become corporate vice president of ITT. He held this position for 14 years. At the end of 1978 he published *Quality Is Free*. This was very well received and in 1979 he left ITT and established Philip Crosby Associates Incorporated, a management consultancy. 1984 saw the publication of another commercially successful book, *Quality Without Tears*.

Philip Crosby retired from Philip Crosby Associates Incorporated in 1991 and founded a company that delivered lecturers and seminars – Career IV Incorporated. In 1996 *Quality Is Still Free* was published and a year later he established Philip Crosby Associates II. The year 1999 saw the publication of his final book *Quality and Me*. Philip Crosby passed away in August 2001.

Crosby's key concepts and contributions to quality theory

Crosby is synonymous with 'zero defects' and a 'do it right first time' approach to quality. He advocates that organisations approach the pursuit of quality in a top down manner, with senior management holding responsibility.

Crosby's Key Concepts on Quality are contained within his:

* Four Absolutes of Quality
* Fourteen Steps to Quality
* Quality Management Maturity Grid
* Five Characteristics Essential to becoming an Eternally Successful Organisation.

He defines 'Four Absolutes of Quality' and a way for implementing organisational improvement – 'Fourteen Steps to Quality'. The Four Absolutes of Quality are:

1   Quality is conformance to requirements.
2   The system for quality is prevention.
3   The performance standard is zero defects.
4   The measurement of quality is the price of non-conformance.

Crosby's Fourteen Steps to Quality are:

1   Make it clear that management is committed to quality.
2   Form quality improvement teams with senior representatives from each department.
3   Measure processes to determine where current and potential quality problems lie.
4   Evaluate the cost of quality and explain its use as a management tool.
5   Raise the quality awareness and personal concern of all employees.
6   Take actions to correct problems identified through previous steps.
7   Establish progress monitoring for the improvement process.
8   Train supervisors to actively carry out their part of the quality improvement programme.
9   Hold a Zero Defects Day to let everyone realise that there has been a change and to reaffirm management commitment.
10 Encourage individuals to establish improvement goals for themselves and their groups.

11 Encourage employees to communicate to management the obstacles they face in attaining their improvement goals.
12 Recognise and appreciate those who participate.
13 Establish quality councils to communicate on a regular basis.
14 Do it all over again to emphasise that the quality improvement programme never ends.

Crosby presents a 'Quality Management Maturity Grid' in his book *Quality Is Free*. This grid serves to provide an organisation with the means to measure its present quality position and is built upon the premise that there are five stages in quality management maturity. These stages are:

1 Uncertainty – management has no knowledge of quality as a positive management tool.
2 Awakening – it is recognised that quality management can help the organisation but no resources are committed.
3 Enlightenment – a decision to introduce a formal quality programme has been made.
4 Wisdom – permanent changes can be made in the organisation.
5 Certainty – quality management is a vital element of organisational management.

According to Crosby there are five characteristics essential to becoming an 'Eternally Successful Organisation'. These are:

1 People routinely do things right the first time.
2 Change is anticipated and used to advantage.
3 Growth is consistent and profitable.
4 New products and services appear when needed.
5 Everyone is happy to work there.

Masaaki Imai

Masaaki Imai was born in Tokyo in 1930 and graduated with a bachelor's degree from the University of Tokyo in 1955. After undertaking graduate work for the University he founded in 1962 The Cambridge Corporation – a consultancy and executive recruitment organisation.

Imai's prominence in the recruitment field was underlined by his 10-year presidency of the Japanese Federation of Recruiting and Employment Agency Associations. His presidency ended in 1986, the year that Imai established the Kaizen Institute – an organisation for the promotion and support of 'Kaizen' concepts. His first book *Kaizen: The Key to Japan's Competitive Success* was also published in the very same year.

The year 1997 saw the publication of Imai's second book about the Kaizen approach to business. This was titled *Gemba Kaizen: A Commonsense, Low-Cost Approach to Management*. Whilst delivering seminars and lectures, Imai presently continues to run the Kaizen Institute.

Imai's key concepts and contributions to quality theory

Masaaki Imai is a proponent of the Kaizen approach to production and has authored two books concerning the topic. So what is Kaizen? Well it is not a single, distinct quality tool – it is an umbrella concept for a number of practices. It is a philosophy of approach to production developed in Japan. According to Imai, Kaizen means "improvement, continuing improvement in personal life, home life, social life, and working life. When applied to the production workplace Kaizen means continuing improvement involving everyone, managers and workers alike". Further to this, Imai suggests that Kaizen is the single most important concept in Japanese management – the key to Japanese success.

The deployment of Kaizen is signified by:

- The evolution of processes through gradual continuous improvement rather than by radical change;
- The recognition of the human resource as the prime company resource; and
- The quantitative measurement of process performance improvement.

Kaizen is not a prescribed method or quality tool but is a continual striving for an incrementally leaner production process that is driven by workplace teams and has improved production process documentation. The management function of Kaizen is considered to be made up of two elements – maintenance and improvement. The 'maintenance' element of the Kaizen approach to production management concerns sustaining current standards through the deployment of policies, rules and standard operating procedures. The 'improvement' element is concerned with incremental improvement – Imai views improvements as being either gradual 'kaizen improvements' or abrupt 'innovations'.

Key features of Kaizen are:

- The empowerment of employees through the use of Kaizen support groups, quality circles and education. People are at the very heart of Kaizen.
- The use of a range of quality tools by employees – including Deming's Cycle and Ishikawa's seven tools – Pareto charts, cause and effects diagrams, stratification, check sheets, histograms, scatter diagrams and control charts.
- The standardisation of workplace processes.
- The undertaking of 'good housekeeping' within the workplace by everyone – using a system known as the '5 S' – to ensure effective work place organisation and continuous incremental improvement. This system involves:

  - 'seiri' – 'sorting out' what is not needed around the individual's workplace,
  - 'seiton' – 'systematically arranging' what is to be kept,
  - 'seiso' – 'scrubbing spick and span' everything that remains,
  - 'seiketsu' – 'spreading and standardising' the routine to others, and
  - 'shitsuke' – 'self-discipline' of establishing a routine schedule for the carrying out of the '5 S'.

- The elimination of 'muda' – this being waste caused by any non-value activity. Eliminating 'muda' creates a leaner, just-in-time production process.

With regard to the Kaizen approach to business practice Imai defines the role of management as being such that they must 'go to gemba'. The word 'gemba' being a Japanese word for 'real place' – the place where the real action happens. In the terms of business activity this place is seen as anywhere that value-adding activities to satisfy the customer are carried out. In the general sense gemba might be where development, production or sales activity takes place. Five principles of gemba-management are presented by Imai:

1   When trouble happens (something abnormal), go to Gemba first.
2   Check with 'gembutsu' (machines, tools, rejects and customer complaints).
3   Take temporary counter-measures on the spot.
4   Find out the root cause.
5   Standardise for prevention of recurrence.

A current leading advocate of Kaizen is Toyota – with a lean, just-in-time approach to production, the creation of a continuous learning culture and the expansion of the employee role.

The Total Quality Management (TQM) approach

The contribution of the seven identified key proponents of quality is significant and each has furthered understanding, development and application of quality management within modern-day organisations. All have contributed in differing ways to the post World War II 'quality revolution' and to various ways of thinking about quality. A key feature of the Western quality revolution of the later part of the twentieth century was the development of a strategic approach to quality management. This approach was labelled 'Total Quality Management'.

Defining Total Quality Management (TQM)

Total Quality Management (TQM) is a management approach, centred on quality, based on the participation of all members and aiming at long-term success through customer satisfaction BSI 1995 (cited in McCabe 2001).

However, since the concept of quality consists of both qualitative and quantitative aspects, quality cannot be directly measured; its assessment contains an element of subjectivity.

Smith (1993) has established four specific factors that impact upon the function of 'quality assessment'. They are the determination of user needs, the identification of entity attributes, assessing the entity's merit on each of the associated attributes and consolidating the established partial scores into a final judgement of quality.

A further aspect requiring consideration with regard to quality is the concept of distinguishing between 'quality' and 'grade'. 'Grade' may be defined as a category or rank given to entities having the same functional use but different technical characteristics. It is worth noting that low quality is usually a problem, but low grade may not be. *Table 1.7* provides an illustration of 'grade' and 'quality' (Project Management Institute 2000).

TABLE 1.7 Grade and quality defined

Software Product (1)	High quality (no obvious bugs, readable manual) and low grade
Software Product (2)	Low quality (many bugs, poorly organised user documentation) and high grade (numerous features)

Quality definitions for TQM

1 **Quality Policy**: policy includes the quality objectives, level of quality required by the organisation, and the allocated roles for organisational employees in carrying out policy and ensuring quality. It shall be supported and implemented by senior management.
2 **Quality Objectives**: objectives are a critical component of the quality policy. For example a quality objective could be to ensure the training of all employees on the quality policy and objectives of the host organisation.
3 **Quality Assurance**: Kerzner (2001) defined Quality Assurance as a "collective term for the formal activities and managerial processes that are planned and undertaken in an attempt to ensure that products and services are delivered at the required quality level".
4 **Quality Control**: Quality Control can be defined as "a collective term for activities and techniques, within the process, that are intended to create specific quality characteristics". In other words, it will assure that the organisation's quality objectives are being met, by using certain techniques such as continually monitoring processes and statistical process control. (Kerzner 2001)
5 **Quality Audit**: Kerzner (2001) opined that it is "an independent evaluation performed by qualified personnel that ensures that the project is conforming to the project's quality requirements and is following the established quality producers and policies".
6 **Quality Plan**: Project team members will create a specific quality plan for the delivery of a specific project. The plan should contain the key elements/ activities of the project and explain in sufficient detail exactly how they are to be delivered and assured.

Smith (1993) addressed the specific issue of how to define quality within a framework of TQM, he suggested it should incorporate two main features:

1 Quality should be taken as the goodness or excellence of organisational products, processes, structures and other entities that an organisation consists of.
2 Quality should be assessed against accepted standards of merit and focus on the requirements of stakeholders.

Accordingly, Quality for TQM purposes can be defined as:

> the goodness or excellence of any product, process, structure or other thing that an organisation consists of or creates. It is assessed against accepted standards

of merit for such things and against the interests/needs of products, consumers and other stakeholders.

(Smith 1993)

Erridge et al. (1998) advocated that the main incentive behind adopting quality initiatives in the UK public sector was attributable to the success of TQM in the private sector. Furthermore, government initiatives have encouraged the application of quality management, noting that it could increase the standards of services being offered by public sector organisations without any corresponding increase in public spending. Moreover, in 1999, the UK government published a white paper entitled 'Modernising Government' which consisted of five commitments:

* To deliver policies that achieve outcomes that matter;
* to deliver responsive public services that meet the needs of citizens, not the convenience of the service provider;
* to deliver efficient, high-quality services and not tolerate mediocrity in service provision;
* to be proactive in the use of new technology in order to meet the needs of citizens and business, and not trail behind technological developments; and
* to value public service and not denigrate it.

The above publication was generated by the Modernising Government Quality Schemes Task Force, which was established in January 1999 by the Cabinet Office. The Cabinet Office led the Task Force with members drawn from across government and organisations managing quality schemes, such as the British Quality Foundation and Investors in People UK.

Another important step was taken towards the improvement of service provision in public sector organisations by establishing *the* Centre for Management and Policy Studies (CMPS). Its purpose is to work with government departments and others in a drive to modernise government. It has been working with the Civil Service College, in order to assess and review the training of civil servants and this has resulted in the creation of new programmes.

Furthermore, Capon et al. (1995) briefly summarised the history of how quality was viewed during the last century. Quality has been measured by the percentage of failures. Then, as prevention and Quality Assurance (QA) became more prevalent, statistical process control and procedural audits provided key measures of its effectiveness. In the 1980s, with cultural change encouraged in a drive for continuous improvement in manufacturing and service provision, employee attitude surveys became popular. In the 1990s, the holistic nature of TQM was adopted, which has encouraged customers', shareholders' and competitors' reactions to become part of the assessment process when assessing the effectiveness of a TQM venture.

Most projects have the conflicting criteria of time, cost and quality. People have differing expectations of quality and these expectations compete with the criteria of cost and time. On this subject Woodward (1997) suggests that time, cost and quality are the prerequisite objectives of any project, but these are not 'compatible' and compromises must be made to try and find the best criteria that fit a particular situation (Woodward 1997). However, projects, be they manufacturing or service centred, are delivered by people. Projects that involve human concerns will raise particularly sensitive issues, and how these issues are dealt with will affect the project's outcome and hence the consideration given to the criteria of time, cost and quality. Further, McGeorge and Palmer (2002) recommended three approaches for considering the relationship between quality and cost:

1   "Higher quality means higher cost"

If a higher standard of quality is required, this usually results in higher costs. In such cases the benefits obtained should be at least equal to the additional cost paid to get the high standard of quality.

2   "The cost of improving quality is less than the resulting savings"

Sometimes during the design stage extra costs have to be incurred in order to improve project quality; this should result in less costs being incurred over the life of the project.

3   "Right-first-time approach"

The costs associated with "not getting it right first time" are higher than the associated costs of "getting it right first time", thus investment in getting it right first time is a worthwhile investment.

The above provides some criteria for engaging in the decision-making process associated with the time, cost and quality dilemma.

Haigh and Morris (2001) noted that the most common difficulty organisations encounter when embarking on the deployment of TQM is the variety of definitions of TQM. A widely recognised definition of TQM has been provided by ISO 8402 (BSI 1995) (formally BS 4778 part 3 1991) which is "a management philosophy embracing all activities through which the needs and expectations of the customer and the community, and the objectives of the organisation are satisfied in the most efficient way by maximising the potential of all employees in a continuing drive for improvement" – while the British Quality Foundation (1998) categorises the definitions of quality into three different types as follows:

1   "Soft aspects", which are concerned with culture, customer orientation, teamwork, and employee participation.

2   "Hard aspects", which are mainly technical aspects such as methods, control of work and procedures.

3   "Soft and hard aspects" address both the technical and humanistic aspects of TQM.

Accordingly, Haigh and Morris (2001) tried to simplify the concept of TQM:

- TQM is a total system of quality improvements with decision making based on facts rather than feeling.
- TQM is not only about the quality of the specific product or service but it is also about everything an organisation does internally to achieve continuous performance improvement.

- TQM assumes that quality is the outcome of all activities that take place within an organisation, in which all functions and all employees have to participate in the improvement process. In other words an organisation requires both Quality Systems and a Quality Culture.
- TQM is a way of managing an organisation so that every job and every process is carried out right first time every time. The key to achieving sustainable quality improvement is through the adoption of TQM principles.

In brief, TQM focuses on a systematic approach to optimally utilise all activities in order to achieve improvements. Therefore "the key aspects of TQM are the prevention of defects and an emphasis on quality in design. TQM is the totally integrated effort for gaining a competitive advantage by continuously improving every facet of an organisation's activities" (Ho 1999).

To simplify the meaning of TQM, Ho (1999) proposed a definition for each word that constitutes TQM:

> **Total**: Everyone associated with the firm is involved in continuous improvement (including its customers and suppliers if feasible).
> **Quality**: Customer's expressed and implied requirements are met fully.
> **Management**: Executives are fully committed. Ideally, everyone in the organisation should be committed.

Griffith et al. (2000) summarised the whole process of TQM as the following: "TQM is a philosophy for achieving a never ending improvement through people". Clearly this statement defines the two essential key factors needed during the process of implementing TQM:

1 Continuous improvement
2 People

Haigh and Morris (2001) advocated that organisations need both 'quality systems' and 'quality culture'. In addition, they added that the transition to sustainable quality improvement cannot be achieved except after embracing and implementing the TQM principles. They further advocate that the works of quality gurus such as Deming, Juran and Crosby could form a basis for understanding the principles of TQM. Furthermore the advocated 'best way' to utilise the works of quality gurus is for an organisation beginning its TQM programme to adopt one of the gurus' works. However, during the implementation process the host organisation should develop its own model, one that better fits its specific criteria.

Deploying Total Quality Management

First it should be noted that the deployment of TQM should be predicted on realistic expectations about "what TQM can deliver even those TQM implementations that have delivered good results but below expectations may be perceived to be failures". (Hendricks and Singhal 2001)

Thus realistic target setting is an activity that Senior Managers must give consideration to, and communicate to staff.

The principles of TQM embrace the concept of customer/supplier relationships existing both within companies (between one person or department and another) and between companies. At each of these interfaces there must be a dedication to meeting the stated requirement with perfection being the only accepted objective. Issues to be addressed as principles of TQM are leadership, commitment, total customer satisfaction, continuous improvement, total involvement, training and education, ownership, reward and recognition, error prevention, co-operation and teamwork (Oakland 1993).

Most if not all construction organisations seek to implement TQM as a valid means of obtaining for their respective organisations a truly sustainable competitive advantage. Competitive advantage has been defined as:

> an advantage your competitors do not have.
>
> *(Hardy 1983)*

Powell (1995) showed that under the resources model, success derives from utilising economically valuable resources that other firms cannot imitate, and for which no equivalent substitute exists. Quality Management can improve a firm's competitiveness through co-operation. Cherkasky (1992) stated that when quality concepts are applied to every decision, transaction and business process, quality becomes a competitive weapon. However, processes which have the greatest impact on customer satisfaction would have to be targeted for improvement and only market research would identify the "key customer drivers" or those products and service attributes of greatest concern to customers.

Chapman et al. (1997) argued that although there was a perception that a quality driven strategic advantage had a direct link with increased business performance, the latter had been difficult to achieve without the development and implementation of a TQM philosophy. Chien et al. (1999) highlighted the factors related to competitive advantage; he identified the following sub-headings: Manufacturing, Marketing, R&D and Engineering and Management.

Fahy (1996) contended that competitive advantage for service firms lies in the unique resources and capabilities possessed by the firm. Not all resources or capabilities are a source of competitive advantage. Only those that meet the stringent conditions of value, rareness, immobility and barriers to imitation are true sources. The actual sources of competitive advantage are likely to vary depending on the nature of the service, the particular traits of the firm, the nature of the industry and the country of origin. *Figure 1.10* provides a pictorial representation of the interrelationship of the sources of competitive advantage, positions of competitive advantage and performance outcomes (Day 1990)

Fahy (1996) concluded that service firms should seek to identify the skills and resources they possess and that they must satisfy the above criteria in order to realise a sustainable competitive advantage.

The linkages between a quality strategy and competitive advantage, though pursued by construction organisations, are very rarely understood within the

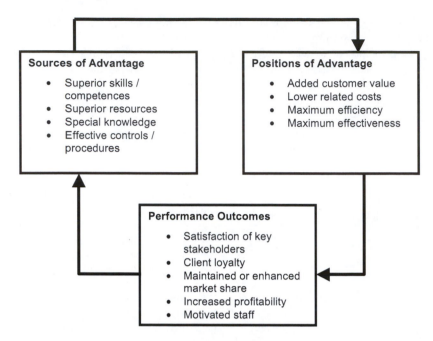

FIGURE 1.10 Sources of sustainable competitive advantage

organisations involved. Improving competitiveness is one of the primary goals of quality management (Rao et al. 1997). Therefore firms need to identify their sources of competitive advantage in order to fully satisfy their clientele.

The linkage between TQM and competitive advantage

From research conducted it has been seen that organisations implementing TQM demonstrate improvements in their efficiency and effectiveness. In the words of one organisation, pursuing TQM had resulted in their being asked to tender for more contracts.

During the onset of the 1992 recession in Australia, major problems arose. Hoffman (1992) identified these as the economy, government reforms, interest rates and the lending market, shortage of labour and lobby group pressures. Hoffman further pointed out that while some companies had 'gone to the wall', others had profited, improved and gained in strength during the same period. His study dealt with the positive elements common to those companies that had profited. The common element was TQM. This verified the hypothesis that TQM improved the efficiency and effectiveness of an organisation. Oakland (1993), as cited by Ghobadian and Gallear (1996), reported the results of a study that compared the performance of 29 companies practicing TQM, along seven key financial indicators for a five-year period, with a corresponding industry median. The study showed that the performance of all the companies that had adopted TQM exceeded their respective industry's median performance level.

Porter's (1990) framework for the analysis of competition in specific industries showed that an industry had a high level of competitive rivalry when:

1   it is easy to enter the market place;
2   both buyers and suppliers had a bargaining power; and
3   there is a threat of substitute products/services entering the market place.

Although Porter's analysis of competitive forces did not specifically address TQM, it does provide a framework for establishing the role that TQM could play in a company's competitive strategy.

The structural implications of TQM for service and manufacturing organisations can be addressed by asking the following key questions:

a   Can TQM be utilised to build barriers against new entrants to the industry?

The barriers of entry are largely dependent upon the size of the organisation. Small and medium sized organisations may gain entry into markets, they are however likely to face competition from other smaller firms wishing to become suppliers to larger organisations. This is due to the increasing demand for a higher quality of service from large organisations (Ghobadian and Gallear 1996). TQM could provide a barrier if clients insisted that it be a pre-requisite before awarding contracts.

b   Can TQM change the basis of competition?

Competition is no longer just between firms from the same sector but also now within a global economy from different sectors. Mohrmam et al. (1995) established a positive correlation between various market conditions and the application of TQM practices. The practices included organisational approaches such as quality improvement teams, quality councils, cross-functional planning, self-inspection, direct employee exposure to customers, collaboration with suppliers on quality efforts, just-in-time deliveries and work cells. Various improvement tools such as the use of statistical process control techniques by front-line employees, process simplification and re-engineering were also evidenced. Measurement systems such as customer satisfaction and cost of quality monitoring also played a vital part. Their studies showed that companies experiencing foreign competition and extreme performance pressures were more likely to use most of the TQM practices, tools and systems. This, they suggested, provided evidence that competitive pressures had led to the adoption of TQM. Betts and Ofori (1992) argued that as trade barriers came down, enterprises in each country would face real competition from firms from other countries, even for small projects.

c   Can TQM change the balance of power in supplier relationships?

Many companies in the manufacturing industry ensure the quality of their products by requiring suppliers to adopt TQM programs (Powell 1995 and Matthews and Burati 1989). Ghobadian and Gallear (1996) corroborated that small and medium sized enterprises (SMEs) were often suppliers of goods and services to larger organisations and in order to remain competitive, they would have to consider the application of TQM due to the increasing demand for higher quality from the larger organisations.

Moreno-Luzon (1993) identified other factors influencing the application of TQM in small and medium sized firms as the pressure of costs, increasing competition, and more demanding customers requiring small firms to implement TQM. TQM works by inspiring employees at every level to continuously improve what they do, thus rooting out unnecessary costs. The competitive advantage results from concentrating resources on controlling costs and improving customer service (both internal and external). Dean (1995) purported the challenge to obtaining a sustainable competitive advantage as being able to holistically define the nature of quality and then rigorously implement a form of integrated product and process development (IPPD) which would attain the defined quality.

TQM enables a company to fully identify the extent of its operational activities and focus them on customer satisfaction. Part of this service focus is the provision of a significant reduction in costs through the elimination of poor quality in the overall manufacturing/service process.

The identified characteristics for a TQM company are essential for it to be able to operate both efficiently and effectively in a dynamic and turbulent environment. Firms require variety in their approach, and hierarchical authoritarian organisations are poorly equipped to provide such a variety. Only business organisations based on the TQM model with vastly reduced bureaucratic control and a rich array of horizontal communication channels, and in which personnel are given a substantial share of authority to make choices and to develop new ideas, can survive under new global market conditions. Adopting a TQM culture takes a substantial amount of time and effort to achieve.

Problematic issues of TQM deployment

Some TQM proponents maintain that a common error in the application of TQM is the failure to recognise that every company, and environment, is different (Laza and Wheaton 1990 cited by Spencer 1994). Successful deployment is considered to be dependent upon the correct alignment of corporate strategies and operational environments with the culture of the host organisation.

A number of problematic issues are commonly associated with the application of TQM; these include:

- **Insufficient commitment by senior management**. Senior management must instil in all employees of the host organisation a desire to improve the competitiveness of the company. TQM's three vital elements are systems, people and resources. Successful implementation is dependent upon senior management developing and organising these key elements. Oakland (1993) advocated that "TQM requires total commitment, which must be extended to all employees at all levels and in all departments". Therefore senior management must be fully committed to the implementation processes. This can be evidenced by senior management providing all resources required for the TQM initiative.

- **Incorrect corporate culture**. TQM requires a corporate culture based on trust and a desire to identify problems in order to eliminate them, thus improving production/service process provision. The concept of 'empowerment' is a vital part of the TQM philosophy. However, if a climate of distrust exists between senior management and other parts of the organisation, the implementation process is doomed to fail. Organisations must understand that a truly 'morphogenic' change is necessary for TQM success and that a cosmetic 'morphostatic' change will not sustain TQM. Organisational culture clearly influences the manner in which a business organisation operates, as well as how employees respond to TQM. As such, an organisation's mission statement must recognise the organisation's culture when drawing up tangible targets that are bounded by closed objectives.

- **No formal implementation strategy**. The implementation process should be planned. TQM is a project and therefore requires planning as a project; to treat it as an organisational bolt-on activity will lead to failure. TQM is a means of improving the competitiveness, effectiveness and flexibility of an entire company. Achieving these noted advantages requires them to plan and organise every operational activity at all levels. This process must be part of the strategic implementation development and should not be treated in isolation. Senior management must also understand that the benefits of implementation are not instantaneous; TQM is a long-term corporate investment and having realistic expectations is vital.

- **Too narrowly based training**. The key to a successful TQM implementation is having staff that are competent to execute their allocated tasks. If employees are empowered to plan and perform work activities it is vital that they also possess the necessary skills and competencies to complete set tasks. A primary function for enterprises seeking to gain a competitive advantage is to implement some form of training initiative which ensures staff have the necessary skills. For example, if staff are to participate in group discussions training in group dynamics and public speaking would be beneficial.

- **Lack of effective communication system**. The life blood of any organisation is communication and the importance of this activity cannot be overemphasised. Within a TQM framework all employees of the company should be able to communicate as necessary and not forget the concept of 'internal' and 'external' customers, with its requirement for effective communication mechanisms. If employees are to become part of the organisational decision-making process, they need a means of expressing their views to senior management. Control within any organisation is dependent upon the effectiveness of the communication systems function.

- **Not concentrating on organisational strengths**. TQM is designed to provide a competitive advantage based upon the host firm's strengths. Senior management should not lose sight of the fact that sustained competitive advantages are obtained by implementing strategies that exploit their strengths through responding to environmental opportunities, while neutralising external threats and avoiding internal weaknesses (Barney 1991). The following two standard corporate planning techniques can be utilised: first a Strengths, Weakness, Opportunities, and Threats analysis (SWOT), and, second, a Political Legal, Economic, Social Cultural and Technological analysis (PEST).

The approach to implementing TQM

Figure 1.11 presents a model that can be adopted or adapted by organisations in order to assist with the development and deployment of TQM approach.

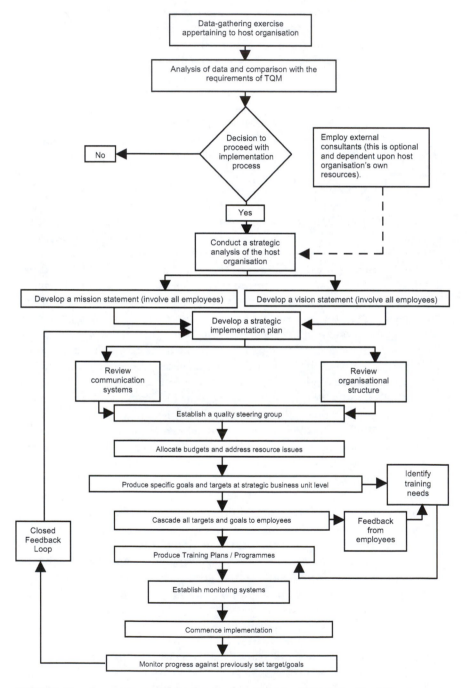

FIGURE 1.11 A generic model for the implementation of Total Quality Management

Management's role in the application of TQM is to create a vision that incorporates TQM as an integral part of the business. Management should further establish organisational policies, structures, and practices consistent with that vision (Ginnodo 1992; Sholter & Hacquebord 1980 cited by Spencer 1994). Managers should be responsible for synthesising all of the different processes and people into a cohesive system (Shores 1992, cited by Spencer 1994).

Thus managers must have a complete understanding of where the company is now, if it wants to deploy TQM and gain the advocated advantages and thus become a 'best practice' organisation.

This will require the company to implement a 'strategic analysis'. Johnson and Scholes (1993) advocate that the strategic analysis should encapsulate "the environment, organisational culture, strategic capability and stakeholder expectations".

Once the organisation attains a full comprehension of its current state, it then needs to establish where it is in relation to its competitors. This process involves benchmarking various activities of the host organisation.

Advocated solutions to problematic issues of TQM deployment

A number of solutions can be proactively proffered to alleviate problems commonly encountered when deploying a TQM approach. These include:

- Senior management must attain a full understanding of the philosophy and requirements of TQM. Senior management is responsible for establishing a quality focused organisation.
- A common vision, one that is recognised and shared by all employees of the organisation. This may be accomplished by adopting awareness sessions, customer surveys, benchmarking and common vision workshops.
- Provision of the necessary resources, which include humanistic as well as financial requirements, education and training for quality improvements.
- The development of a holistic deployment strategy that may be based on an incremental process. Senior management must review the quality management systems in order to monitor and maintain progress.
- Design procedural systems relating to work practices. Concentration of organisational effort should be placed on prevention rather than corrective actions.
- Organisations must effectively balance between processes and results. Some organisations focus only on processes and neglect the importance of results.

Factors that influence TQM implementation in public sector organisations

Dewhurst et al. (1999) propose that there are key dimensions that influence TQM implementation within public sector organisations. These dimensions are:

- Top management support: commitment and leadership of top management.
- Customer relationship: culture change, customer involvement, and customer focus.

- Culture change: training, employee empowerment, education and employee relations,
- Supplier relationship: culture change, supplier quality management and supplier involvement.
- Workforce: management and teamwork.
- Employee attitudes and behaviour: employee involvement.
- Product and/or service design process: product/service design and design quality management.
- Process flow management: use of tools and techniques and process management/operating procedures.
- Quality data and reporting: measurement and feedback, quality data and reporting and internal quality information.
- Role of the quality department: it should play an integral part and ensure it leads on the initiative.
- Benchmarking as a means of self-assessment.

The advantages of applying a TQM approach

TQM can be advocated as a solution for organisations that are underperforming due to their use of traditional organisation structures and management practices whilst operating within a dynamic environment. The implementation of a TQM philosophy can facilitate performance in such organisations.

The advantages of applying a TQM approach are:

- the production of a higher quality product/service through the systematic consideration of client's requirements;
- a reduction in the overall process/time and costs via the minimisation of potential causes of errors and corrective actions;
- increased efficiency and effectiveness of all personnel with activities focused on customer satisfaction; and
- improvement in information flow between all participants through team building and proactive management strategies.

TQM can assist in making effective use of all organisational resources, by developing a culture of continuous improvement. This empowers senior management to maximise their value-added activities and minimise efforts/organisational energy expended on non-value-adding activities.

TQM enables companies to fully identify the extent of their operational activities and focus them on customer satisfaction. Part of this service focus is the provision of a significant reduction in costs through the elimination of poor quality in the overall process. This empowers companies to attain a truly sustainable competitive advantage. TQM provides a holistic framework for the operational activities of enterprises. If a firm can overcome the problematic issues of implementation, then a sustained competitive advantage is the reward to be gained.

A tool for assisting management and employees to define and comprehend TQM within their own organisation is the 'European Foundation for Quality Management Excellence Model' (EFQM Model).

Summary

This chapter has presented various definitions and notions of quality. It has introduced people and theories that have contributed significantly to the development of the concept and practice of quality management in modern-day organisations. This chapter has also considered TQM and the advantages and problematic issues associated with implementing a TQM approach within a modern-day organisation.

Questions for the reader

Here follows a number of questions related specifically to the information presented within this chapter. Try to attempt each question without reference to the chapter in order to assess how much you have learned. The answers are provided at the end of the book.

Question 1

Define the following terms:

1a) Quality policy
1b) Quality objectives
1c) Quality assurance
1d) Quality control
1e) Quality audit
1f) Quality plan

Question 2

The concept of Total Quality Management has been simplified to four aspects (Haigh and Morris 2001). Identify the four aspects of TQM.

Question 3 – case study

You have been asked to act as an external consultant for 'Monaghan and Monaghan Developments' (M&M Developments). M&M Developments are considering the implementation of a formal TQM system with a view to obtaining externally verified ISO accreditations. M&M Developments consider accreditation to be a necessity in order to be placed on tender lists and continuously improve their operations.

As an external consultant, you are requested to prepare and deliver a presentation to the senior partners of 'Monaghan and Monaghan Developments'. The topic of the presentation is 'the benefits of TQM and the associated implementation process'. Prepare notes to facilitate this presentation.

Further reading

Philip Crosby

Crosby, P. (1972). *Situation Management: The Art of Getting Your Own Sweet Way.* New York: McGraw-Hill Education.

Crosby, P. (1978). *Quality Is Free: The Art of Making Quality Certain.* New York: McGraw-Hill Education.

Crosby, P. (1982). *The Art of Getting Your Own Sweet Way.* New York: McGraw-Hill.

Crosby, P. (1984). *Quality Without Tears: The Art of Hassle Free Management.* New York: McGraw-Hill Education.

Crosby, P. (1996). *Quality Is Still Free: Making Quality Certain in Uncertain Times.* McGraw-Hill Education.

Crosby, P. (1999). *Quality and Me: Lessons from an Evolving Life.* San Francisco: Jossey Bass Wiley.

W Edwards Deming

Deming, W.E. (1943). *Statistical Adjustments for Data.* London: Chapman and Hall.

Deming, W.E. (1950). *Some Theory of Sampling.* London: Chapman and Hall.

Deming, W.E. (1960). *Sample Designs in Business Research.* Hoboken, NJ: John Wiley & Sons Inc.

Deming, W.E. (1982). *Quality, Productivity, and Competitive Position.* Cambridge, MA: Massachusetts Institute Technology.

Deming, W.E. (1986). *Out of Crisis.* Cambridge, MA: Massachusetts Institute of Technology.

Deming, W.E. (1993). *The New Economics: For Industry, Government, Education.* Cambridge, MA: Massachusetts Institute of Technology.

Armand V Feigenbaum

Feigenbaum, A.V. (1951). *Quality Control: Principles and Practice.* New York: McGraw-Hill.

Feigenbaum, A.V. (1961). *Total Quality Control.* New York: McGraw-Hill.

Feigenbaum, A.V., and Feigenbaum, D.S. (2003). *The Power of Management Capital.* New York: McGraw-Hill.

Masaaki Imai

Masaaki, I. (1986). *Kaizen: The Key to Japan's Competitive Success.* New York: McGraw-Hill Education.

Masaaki, I. (1997). *Gemba Kaizen: A Commonsense, Low-Cost Approach to Management.* New York: McGraw-Hill Education.

Kaoru Ishikawa

Kaoru, I. (1984). *Guide to Quality Control.* Tokyo: Asian Productivity Organization.

Kaoru, I. (1985). *What Is Total Quality Control? The Japanese Way.* Englewood Cliffs, NJ: Prentice Hall.

Joseph Juran

Juran, J. (1951). *Quality Control Handbook.* New York: McGraw-Hill.

Juran, J. (1955). *Case Studies in Industrial Management.* New York: McGraw-Hill.

Juran, J. (1964). *Managerial Breakthrough.* New York: McGraw-Hill.

Juran, J., & Frank, G. (1970). *Quality Planning and Analysis.* New York: McGraw-Hill Education.

Juran, J. (1982). *Juran on Quality Improvement.* New York: Juran Institute.

Juran, J. (1988). *Juran on Planning for Quality.* New York: The Free Press.

Juran, J. (1998). *Juran on Quality By Design: The New Steps for Planning Quality into Goods and Services: Planning, Setting and Reaching Quality Goals.* New York: Simon & Schuster Inc.

Genichi Taguchi

Taguchi, G. (1986). *Introduction to Quality Engineering: Designing Quality into Products and Processes.* Tokyo: Asian Productivity Organization.

Taguchi, G. (1994). *Taguchi methods: On-Line Production* (Quality Engineering Series). Illinois: Irwin Professional Publishing.

Taguchi, G. (2001). *An Introduction to Quality Engineering.* Tokyo: Asian Productivity Organisation.

References

Barney, J. (1991). Firm resources and sustained competitive advantage. *Journal of Management*, 17 (1), pp. 99–120.

Betts, M., and Ofori, G. (1992). Strategic planning for competitive advantage in construction. *Construction Management and Economics*, 10, pp. 511–532.

British Quality Foundation (1998). *Self Assessment Techniques for Business Excellence: Identifying Business Opportunities.* London.

British Standards Institution (1995). *Quality Management and Quality Assurance Vocabulary.* BS EN ISO 8402 (formally BS 4778: Part 1, 1987/ISO 8402, 1986). London.

Capon, N., Kay, M., and Wood, M. (1995). Measuring the success of a TQM programme. *International Journal of Quality & Reliability Management*, 12 (8), pp. 8–22.

Chapman, R.L., and Murray, P.C., and Mellor, P. (1997). Strategic quality management & financial performance indicators. *International Journal of Quality & Reliability Management*, 14 (4), pp. 432–448.

Cherkasky, S.M. (1992). Total quality for a sustainable competitive advantage. *Quality*, 31 (8), Q4.

Chien, T.W., Lin, C., Tan, B., and Lee, W.C. (1999). A neural networks-based approach for strategic planning. *Information and Management*, 35, pp. 357–364.

Crosby, P. (1979). *Quality Is Free: The Art of Making Quality Certain.* New York: McGraw-Hill Education.

Day, S.G. (1990). *Market Driven Strategy: Process for Creating Value.* New York: The Free Press, p. 128.

Dean, E.B. (1995). Total Quality Management from the perspective of competitive advantage. NASA Langley Research Center. Available at: http://spartan.ac.brocku.ca/~pscarbrough/dfca1stmods/dfc/tqm.html. [Accessed 10 August 2017].

Deming, W.E. (1986). *Out of Crisis.* Cambridge, MA: MIT Press.

Dewhurst, F., Martinez-Lorente, A., and Dale, B. (1999). TQM in public organisations: An examination of the issues. *Managing Service Quality*, 9 (4), pp. 265–273.

Erridge, A., Fee, R., and McIlroy, J. (1998). Public sector quality: Political project or legitimate goal? *International Journal of Public Sector Management*, 11 (5), pp. 341–353.

Fahy, J. (1996). Competitive advantage in international services: A resource-based view. *International Studies of Management & Organisation*, 26 (2), pp. 24–37.

Feigenbaum, A.V. (1951). *Quality Control: Principles and Practice*. New York: McGraw-Hill.

Feigenbaum, A.V. (1961). *Total Quality Control*. New York: McGraw-Hill.

Garvin, D. (1986). Quality problems, policies and attitudes in the US and Japan: An exploratory study. *Academy of Management Journal*, 29 (4), pp. 653–673.

Garvin, D. (1988). *Managing Quality*. New York: The Free Press.

Ghobadian, A., and Gallear, D.N. (1996). Total Quality Management in SMEs. *Omega International Journal of Management Science*, 24 (1), pp. 83–106.

Ginnodo, W. L. (1992). How TQM is defining management and leadership. *Tapping the Network Journal*, 3 (3), p. 8.

Gitlow, H., Gitlow, S., Oppenheim, H., and Oppenheim, R. (1989). *Tool and Methods for the Improvement of Quality*. Homewood, IL: Irwin.

Griffith, A., Stephenson, P., and Watson, P. (2000). *Management Systems for Construction*. 1st ed. London: Longman.

Haigh, B., and Morris, D. (2001). *Total Quality Management, a Case Study Approach*. 1st ed. Sheffield: Sheffield Hallam University Press.

Hardy, L. (1988). *Successful Business Strategy: How to Win the Market Place*. London: Kogan Page.

Hendricks, K.B., and Singhal, V.R. (2001). Firm characteristics, Total Quality Management, and financial performance. *Journal of Operations Management*, 19, pp. 269–285.

Ho, S. (1999). From TQM to business excellence. *Production Planning and Control*, 10 (1), pp. 87–96.

Hoffman, K. (1992). *Improving Business Performance in the Building Industry through Total Quality Management*. Conference proceedings for ABIC '92 on efficient & effective construction in the 90's, Australia, Gold Coast.

Imai, M. (1986). *The Key to Japan's Competitive Success*. McGraw-Hill, New York, NY.

Johnson, G., and Scholes, K. (1993). *Exploring Corporate Strategy*. London: Prentice Hall.

Joiner, B.L. (1993). *Fourth Generation Management*. New York: McGraw-Hill Education.

Juran, J. (1988). *Juran on Planning for Quality*. New York: The Free Press.

Kerzner, H. (2001). *Project Management: A Systems Approach to Planning, Scheduling, and Controlling*. 7th ed. Hoboken, NJ: John Wiley & Sons, Inc.

Masaaki, I. (1986). *Kaizen: The Key to Japan's Competitive Success*. New York: McGraw-Hill Education.

Matthews, M.F., and Burati, J.L. Jr. (1989). *Quality Management Organisations and Techniques*. Source Document 51. Austin, Texas: The Construction Industry Institute.

McCabe, S. (2001). *Benchmarking in Construction*. Oxon, UK: Blackwell Science.

McGeorge, D., and Palmer, A. (2002). *Construction Management: New Directions*. 2nd ed. Oxford: Blackwell Science Ltd.

Mohrmam, S.A., Tenkasi, R.V., Lawler III, E.E., and Ledord, G.E. Jr. (1995). Total Quality Management: Practice and outcomes in the largest US firms. *Employee Relations*, 17 (3), pp. 26–41.

Moreno-Luzon, M.D. (1993). Can Total Quality Management make small firms competitive? *Total Quality Management*, 4 (2), pp. 165–181.

Oakland, S.J. (1993). *Total Quality Management*. London: Butterworth Heineman.

Parasuraman, A., Zeithaml, V.A., and Berry, L.L., (1988). Servqual: A multiple-item scale for measuring consumer perceptions of service quality. *Journal of Retailing*, 64, pp. 12–40.

Porter, M.E. (1990). The competitive advantage of nations. *Harvard Business Review*, March–April, 90 (2), pp. 73–95.

Powell, T.C. (1995). Total Quality Management as competitive advantage: A review and empirical study. *Strategic Management Journal*, 16, pp. 15–37.

Project Management Institute (2000). *Guide to the Project Management Body of Knowledge* (PMBOK Guide). Newtown Square, PA: PMI.

Rao, S.S., Ragu-Nathan, T.S., and Slis, L.E. (1997). Does ISO 9000 have an effect on quality management practices? An international empirical study. *Total Quality Management*, 8 (6), pp. 335–346.

Reeves, C.A., and Bednar, D.A. (1994). *Academy of Management Review*, 19 (3), Special Issue: "Total Quality" (July, 1994), pp. 419–445.

Shewhart, W.A. (1931). *Economic Control of Quality of Manufactured Product*. New York: van Nostrand.

Smith, G. (1993). The meaning of quality. *Total Quality Management*, 4 (3), pp. 235–245.

Spencer, B.A. (1994). Models of organisational and Total Quality Management: A comparison and critical evaluation. *Academy of Management Review*, 19 (3), pp. 446–471.

Taguchi, G. (1986). *Introduction to Quality Engineering: Designing Quality into Products and Processes*. White Plains, NY: Quality Resources. Asian Productivity Resources, Quality Resources.

Woodward, J. (1997). *Construction Project Management, Getting It Right First Time*. 1st ed. London: Thomas Telford Services Ltd.

Zhang, Q. (2001). Quality dimensions, perspectives and practices: A mapping analysis. *International Journal of Quality & Reliability Management*, 18 (7), pp. 708–721.

2

MEASURING PROJECT AND CORPORATE PERFORMANCE

This chapter provides a concise introduction to the concepts of stakeholders and self-assessment. These are two key concepts and components in the pursuit of project and corporate quality and improvement. Whilst a construction enterprise needs to understand who its stakeholders are, and what is required by them, self-assessment serves to enable a construction company to determine if it is in fact meeting the requirements of their stakeholders. This chapter highlights concepts and practices relating to stakeholders, key performance indicators (KPIs) and benchmarking. These are presented in an introductory form as discussion and application of self-assessment is further detailed and extended within Chapters 4 and 6 of this text.

After all a (construction) business must change to stay ahead or get ahead. If a business does not keep up then its only option is to fall behind (MacDonald and Turner 1998).

Learning outcomes

By the end of this chapter the reader will be able to demonstrate an understanding of:

- The fundamental principles of the stakeholder concept.
- The importance of conducting project and organisation self-assessment activity.
- The benefits of looking externally as well as internally, when engaging in self-assessment activity.

Stakeholders

The original concept of stakeholders related to enterprises and encapsulated groups or individuals who had a financial 'stake' in the corporation. This original concept has now been expanded to incorporate the true concept of 'stakeholder theory'.

Stakeholder theory relates to project and corporate organisational management and business ethics and it addresses both morals and values in managing a construction related enterprise. A critical and evaluative approach to stakeholders was provided by Freeman (1984). Freeman's work was related to the identification and modelling of groups that could be classified as the stakeholders of companies. Freeman advocated a methodology that can be adopted by management. This methodology addressed the sometimes disparate approach taken by organisations, including construction companies, in identifying stakeholders and their requirements. Freeman put the stakeholder concept in very simplistic terms; it addresses the 'Principle of Who or What really counts' (Freeman 1984).

One needs to give careful consideration not only as to who are an organisation's stakeholders, but also, consideration must be given as to what influence they can exert upon the host company. Mitchell et al. (1997) produced a typology of stakeholders; this work was based on the attributes of power, or the extent that they could influence a company. It is therefore a wise strategy for construction companies to be aware of the stakeholder concept, and the impact its stakeholders could exert upon them.

A construction firm must effectively manage its relationship with its stakeholders, but first it needs to identify its stakeholders. The following provides some examples of typical stakeholders related to their respective interests:

•	Government	Taxation, legislation, lower unemployment
•	Senior management	Performance issues, target setting, corporate growth and morale, ethical/green issues and corporate longevity
•	Non-managerial staff	Pay rates, job security, morale, facilities
•	Unions	Working conditions, pay rates, legal requests, health and safety issues
•	Customers	Value for money, quality, customer care, ethical products
•	Creditors/suppliers	Liquidity, timely payments, new contracts, corporate longevity
•	Local communities	Jobs, local investment, environmental issues, ethical practices

The type of stakeholders along with their respective interests and influences will vary from company to company and industry to industry.

Typical stakeholders

- People who will be affected by an endeavour and can influence it, but who are not directly involved with doing the work. Any group or individual who can affect or who is affected by achievement of a group's objectives;
- Any individual or group with an interest in a group's or organisation's success in delivering intended results and in maintaining the viability of the group or

the organisation's product and/or service. Stakeholders influence programmes, products and services;
- Any organisation, governmental entity or individual that has a stake in or may be impacted by a given approach to environmental regulation, pollution prevention, energy conservation, etc;
- A participant in a community mobilisation effort, representing a particular segment of society.

Stakeholders that are classed as 'primary' are those that engage in economic business activity with the host construction company, for example, suppliers and subcontractors.

Stakeholders that are classified as 'secondary' are those who – although they do not engage in direct business activity – are affected by or can affect its actions. For example the media, via adverse press related to a construction company or one of its projects.

So the term 'stakeholder' has become more commonly used to mean a person, group or organisation that has a legitimate interest in a project or company. When considering the decision-making processes for enterprises, or government agencies, and non-profit organisations, the concept has been extended to encompass all those who have an interest (or 'stake') in what the enterprise does. This includes not only its vendors, employers, and customers, but even members of a community where its offices or projects may affect the local economy or environment.

Stakeholder mapping

The production of a stakeholder map is a very useful activity for construction firms (or any firm) to undertake. The underpinning rationale for the production of a stakeholder map is that:

- It is used to identify all interested parties, both inside and outside the business. Remember, any organisation has both internal and external stakeholders.
- It is used to ensure that stakeholder interests are established and catered to.
- It can assist in balancing the needs/interests of the various stakeholders.

Stakeholder content map

A stakeholder map in pictorial form is presented in *Figure 2.1*. The production of a map is a very good starting point when considering the interconnection of internal and external stakeholder relationships.

Stakeholder mapping is an essential activity for construction companies to engage in. It is essential because if a construction company does not know who its stakeholders are, or what they require, how can that construction company possibly meet their requirements?

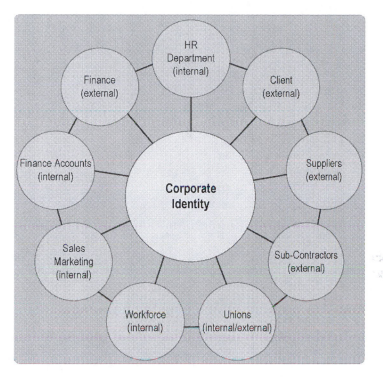

FIGURE 2.1   Typical example of a stakeholder map

The development of key performance indicators

Construction companies, or any company for that matter, need to be able to gauge their performance; this is essential if that company is striving for organisational improvement, via organisational learning (see Chapter 6 of this text book).

Thus the concept of Key Performance Indicators (KPIs) has been embraced by the UK Construction Industry. The KPIs scheme for construction firms evolved following the report *Rethinking Construction*. The Construction Industry Task Force prepared this report in 1998 for the Department of the Environment, Transport and the Regions (DETR). The report highlighted a number of areas where the construction industry could improve its performance.

One of the fundamental reasons why the construction industry needed to institute change arose from the statement made in the report:

> To drive dramatic performance improvement the task force believes that the construction industry should set clear measurable objectives and then give them focus by adopting qualified targets, milestones and performance indicators.
>
> *(Construction Industry Task Force 1998)*

The Construction Task Force identified specific targets for improvements, these being:

- Capital Costs reduce by 10 percent
- Construction time reduce by 10 percent
- Predictability reduce by 20 percent
- Defects reduce by 20 percent
- Accidents reduce 20 percent
- Productivity increase by 10 percent
- Turnover and profits increase by 10 percent (McCabe 2001).

Following the production of the *Rethinking Construction* report, the Construction Best Practice Programme (CBPP) was developed by the DETR, the Construction Industry Board (CIB) and the Movement for Innovation (M4I).

The CBPP was established to institute the challenges that were set to the industry made by *Rethinking Construction* amongst several other initiatives, the KPI scheme was launched in May 1999.

KPI Working Group (2000) states:

> The purpose of the KPIs is to enable measurement of project and organisational performance throughout the construction industry. This information can then be used for benchmarking purposes, and will be a key component of any organisation's move towards achieving best practice.

Since the formulation of the initial set of KPIs, other indicators have been established; these include People KPIs, Construction Products KPIs and Environmental KPIs.

The M4I set 10 KPIs, as noted by Cook:

> Seven are applied on a project-by-project basis: Construction cost; Construction time; Cost predictability; Time predictability; Defects; Product satisfaction; and Customer satisfaction. Three indicators look at Company performance; Profitability; Productivity; and Safety.
>
> *(Cook 1999)*

It is intended that KPIs be used throughout the construction industry by all parties, including clients, designers, consultants, contractors and subcontractors. The 10 KPIs are also covered by:

> a set of 'super graphs' that plots the entire industry's average performance for all 10 indicators. There are also five sector-specific sets of graphs covering new-build housing; new-build non-housing; Infrastructure; repair; maintenance housing; and repair, maintenance non-housing.
>
> *(Cook 1999)*

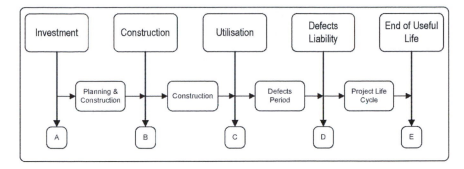

FIGURE 2.2   Process of commitment

The use of the graphs assists construction companies in accurately benchmarking themselves against other construction companies.

As construction covers the design, construction and eventually the demolition of a project, this would enable the same project to be measured using the same set of indicators but with the differing parties involved.

The KPI Working Group set out to define the five key stages of a project and these are depicted in *Figure 2.2*.

Construction key performance indicators

The Key previously stated Performance Indicators are known as the 'headline KPIs'. Other performance indicators are in existence and cover the following areas:

- Operational Indicators
- Diagnostic Indicators

Operational Indicators relate to specific aspects of a construction company's activities and should enable senior management to identify and focus on specific areas for improvements. Diagnostic indicators provide information on why certain changes may have occurred in the headline or operational indicators; they are useful in analysing areas for improvement in more detail. This enables a construction company to efficiently and effectively focus its limited organisational resources.

When engaging in this type of activity a construction company must consider the 'Law of diminishing returns'. There is always an organisational cost to be paid when engaging in 'Change management activities'.

KPIs provide useful tools for engaging in self assessment activities by enabling the measuring and analysing of the company and/or project, in order to obtain an improved level of performance. The headline indicators provide the main focus of performance measurement.

In order to constructively assess the ten headline KPIs it is necessary to identify the primary objectives of each particular project indicator. (The points noted in the following section relate to *Figure 2.2*.)

1. Project indicator: construction cost

> Defined as a change in the current normalised construction cost of a project at Commit to Construct (point B) compared with one year earlier, expressed as a percentage of the one year earlier cost.
>
> *(Great Britain, The KPI Working Group 2000)*

This indicator sets out to measure the construction cost of a project this year and compares it to the cost of a similar project constructed last year. 'Rethinking Construction' set a target to reduce the year on year cost by 10 percent. The method in which to measure the construction cost involves considering 'two identical structures', when each are:

> completed in successive years and the second is finished for 10% less than the first, then the cost indicator would be −10%.
>
> *(Cook 1999)*

Very rarely are two construction projects the same, so in order to establish the cost of two projects, normalisation factors are used to take into account such variables as location, building materials, process and size. The resulting cost indicator is then plotted on the relevant line graph; the benchmark score is then defined. The benchmark score represents the organisation's performance, with the industry standard performance being 50 percent.

2. Project indicator: construction time

The construction time can be defined as:

> A change in the current normalised construction time of a project at Commit to Construct (point B) compared with one year earlier, expressed as a percentage of the one year earlier time.
>
> *(Report of the Minister of Construction 2000)*

It is 'Rethinking Construction's' target to reduce the construction time by 10 percent year on year. Cook (1999) opines that the construction time can be measured by comparing two similar projects that are finished a year apart. If the second project is completed in 10 percent less time after the normalisation factors have been applied, then the time indicator is −10 percent.

3. Project indicator: cost predictability

A problem identified by the 'Rethinking Construction' report was the relative uncertainty of the construction costs. This indicator is defined as the:

> Change between the actual construction cost at Available for Use (point C) and the estimated construction cost at Commit to Construct (point B).
>
> *(Report of the Construction Minister 2000)*

'Rethinking Construction' set the challenge to make an improvement of 20 percent year on year.

4. Project indicator: time predictability

The Time predictability indicator has the same reasoning as the Cost Predictability indicator as there was a need at the time of production of 'Rethinking Construction' to improve the predictability of time during all stages of the project. The Time predictability is the:

> Change between the actual construction time at Available for Use (point C) and the estimated construction time at Commit to Construct (point B), expressed as a percentage of the estimated construction time at Commit to Construct (point B).
>
> *(Report of the Minister of Construction 2000)*

5. Project indicators: defects

The work of Cook (1999) defines the defects indicator as how the handover of the project was affected by defects. Four individual terms may be used in order to classify the defects:

I   Defect free
II   Few defects and available for use on handover
III   One or more defects and slight delay
IV   Major defects that substantially delayed handover

6–7. Project indicators: products and service satisfaction

These two indicators measure the client's satisfaction with the finished product and how satisfied they are with the services they received from the project team. These indicators are measured by asking the client at the end of the project to give the organisation a rating between 0 and 10 for each performance indicator with 10 indicating totally satisfied.

8. Company indicator: profitability

Cook (1999) defines this indicator as a 'company's pre-tax profit as a percentage of sales'; this indicator has a target to increase the company's turnover and profit by 10 percent year on year.

9. Company indicator: productivity

The productivity indicator sets out to establish how much turnover each full time employee generates for the employing organisation. Its 'Rethinking Constructions' target is to improve a company's productivity by 10 percent each year.

10. Company indicator: safety

This indicator strives to measure the number of accidents per 100,000 employees and reduce them by 205 every year.

There are seven basic steps to be followed by construction companies when implementing the ten KPIs. These steps are presented in *Figure 2.3*.

Figure 2.3 illustrates that the KPIs have been designed to continually monitor and review the measures of a construction company's performance. There does exist in the construction industry a view that a lack of appropriate performance measurement does hinder organisational and industry improvement. For continuous improvement to occur it is necessary to have performance measures which check and monitor actual performance, to verify changes and the effect of improvement actions. The EFQM.E.M described in Chapter 4 of this text book provides an excellent model for addressing this critical issue.

The following is a generic process and can be adopted/adapted to the other KPI schemes.

Step 1   Decide what to measure

The construction industry has a multiplicity of KPIs and these can be used to measure the performance of a company or an individual project. There is a danger that people involved in the setting and interpreting of KPIs may be utilising incorrect KPIs, resulting in the production of invalid and misleading data. It is vital that a

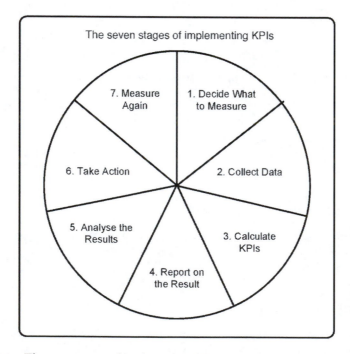

FIGURE 2.3   The seven stages of implementing KPIs

company clearly determines those areas which need to be measured and this must relate to a strategic approach to improvement activities. The selection of appropriate measurement systems and procedures is a very critical activity when undertaking a corporate or project monitoring system. Companies undertake monitoring activities in order to control and evaluate variations and improvements. A construction firm should ensure that it measures what is important to both the firm and its stakeholders.

In the early stages of KPI deployment companies have a tendency to try and measure too many KPIs; this can result in confusion. The best strategy for a firm is to identify those KPIs that are critical to its activities, and commence with those. A good place to start measuring KPIs is to measure the ten standard KPIs, in order to assess the current state of the organisation. It should be noted that the more traditional criteria usually measured in projects, for example costs and schedule times, are not necessarily appropriate for engagement in continuous improvement activities (see Chapter 6 of this text book). This is because they are not completely effective in identifying the root causes of productivity and quality problematic issues. Also they do not provide an adequate vision of the potential for improvement, and information obtained usually arrives too late to take timely effective corrective actions. One should not forget that it is not possible to take effective retrospective corrective actions.

Step 2   Appropriate data collection

Data that need to be collected will initially come from the contractor's existing records, clients and suppliers. As an example, the information required to apply the Health and Safety KPIs needs to be collected from internal records. Whereas the information required for addressing client satisfaction, in terms of product and service levels, needs to come from the client. In order to be able to accurately measure performance, the information should be valid, reliable and timely.

Step 3   Calculation of KPIs

In order to commence the calculation of KPIs a construction company must first decide which set of data it is going to benchmark against. Many construction firms compare themselves against the 'All Construction' data set. This data set contains all key sectors of the construction industry' except material suppliers' however, it may be better for firms to compare themselves against more appropriate KPIs.

Having established a valid methodology for comparing the results, the next stage is to calculate the individual KPIs, and a suitable method would be to follow the rules as defined by the KPI Working Group. The calculated scores can then be plotted on to each individual indicator graph, in order to determine the benchmark score. The benchmark score will be a value between 0 and 100 percent; a value of 50 percent represents the average performance of other sector companies.

Step 4   Reporting the results of the analysis

Resulting individual benchmark scores have to be both Feedback and Fed Forward, if any benefit is to be achieved. The benchmark scores can be plotted on to a radar chart.

Pearson (2002) advocates: "the purpose of Key Performance Indicators is to show how a company's performance compares with the average achieved by the industry – shown by the 50% circle in the radar charts".

Step 5   Analysis of results and implementation

Construction companies must ensure that the resulting analysis is presented in a manner that can be readily understood by those people who are charged with addressing any variances, in the drive for improvements. Too many construction companies stop at the analysis of the results stage and never obtain the full benefit of using their performance data in making continuous corporate and project improvements. Fitting actions should be taken in order to maintain corporate strengths and eliminate/reduce corporate and project weaknesses. KPIs cannot by themselves inform a company what actions need to be taken in order to improve corporate and project performance.

Step 6   Take action

One must remember that the taking of any action has to be made in a timely manner. It is not possible to have effective retrospective corrective actions. Also the actions taken must relate to the issues under consideration. So it is important to correctly establish the cause and effect relationship. Having taken some corrective actions it is then important to gauge their effectiveness.

Step 7   Re-measure

Continuous improvement is a cyclical activity, thus, once the first cycle of measuring KPIs has been undertaken and measures put in place to improve the performance, the noted activities should be measured again. This will ensure that the measures that have been taken actually led to an improvement on the previous results.

The utilisation of KPIs is based on the strategic action of trying to improve and/or gain a sustainable competitive advantage. It is worth remembering that a construction firm's competitive advantage is based on the value it is able to create for its clients, which exceeds the firm's cost of creating it. KPIs can assist in increasing value without increasing corresponding costs for the company and its internal and external stakeholders. When deploying KPIs construction firms should ensure that KPIs relate to their adopted business strategy. A simple check list is to ensure that:

- Companies focus on KPIs that have an immediate benefit to their business activities and clients.
- Companies use KPIs that are valid, easily measured and readily understood.
- People get used to reporting and displaying the resulting data.
- KPIs become an integral part of any corporate and project performance assessment activity.
- KPIs ensure a company develops a continuous improvement culture.

Benchmarking

Simplistically, benchmarking is a form of individual and organisational learning, though such learning has been described as "Adaptive Learning" (see Chapter 5 of this text book). Adaptive learning is more related to 'single-loop learning', with the inference being that it is more akin to copying than actual learning. However, it can be seen to contribute to the continuous striving for improvement in personal, project and corporate performance. Many construction organisations do now encourage and engage in benchmarking activities as a fundamental business practice, some with the aim of becoming 'best in class'.

Defining benchmarking

Many definitions of benchmarking exist, but it is fundamentally concerned with making valid comparisons between other organisations or projects, and then learning the lessons that these comparisons reveal.

The Royal Mail have defined benchmarking as a structured process of learning from others, internally or externally, who are leaders in a field or with whom legitimate comparisons can be made. The American Productivity & Quality Centre (APQC) define it as a process of continuously comparing and measuring against other organisations anywhere in the world to gain information on philosophies, policies, practices and measures which will help our organisation take action to improve its performance.

Benchmarking should not just focus on the obtaining of performance measures, to become truly effective it has to become part of a construction company's core business strategy, in a drive to keep an organisation at the competitive edge. The essential elements of an effective and efficient benchmarking activity are that the practice is:

- Continuous: benchmarking should not be treated as a 'one-off' exercise; it should be incorporated into the regular planning cycle of the construction organisation, and part of the management of key processes. Its true value is in its being part of an iterative continuous improvement process.
- Systematic: it is important for construction firms to ensure that a valid and consistent methodology is adopted, and that it is actually followed. It is also very important that processes are in place to ensure that good practice is shared across the organisation, if the company is to obtain the true benefits of engagement with the benchmarking activity.
- Implementation: benchmarking assists in the identification of any gaps that may exist between an organisation's current performance and that required when performing at 'Best Practice' level; this is achieved by conducting a comparative analysis. A host company must know how 'Best Practice' performance can be achieved, and have the resources to deploy any necessary actions. For improvements to occur at corporate and project levels a clear set of actions need to be established and implemented.
- Best Practice: the identification of absolute 'Best Practice' is not essential for benchmarking to be deemed successful. It is more likely that obtaining a measure of 'Good Practice' would be acceptable.

The European Foundation for Quality Management (EFQM) considers the basic philosophy of benchmarking to be:

- Knowing what you want to improve/learn about (Scoping)
- Identifying the 'Good Practices' in those areas
- Learning from the 'Good Practices' – organisations:

  - What they are achieving
  - How they are achieving it

- Adapt the Key Insights and incorporate the learning into your own processes.

Chapter 4 of this text book provides further detailed information on this topic.

In summarising, it may be stated that benchmarking empowers construction companies to adopt/adapt and improve organisational practices. The engagement of benchmarking requires a significant focus to be placed on 'process thinking'. This requires an organisation to realise that it consists of a set of processes that are cross-cutting. A process can be defined as a sequence of activities that adds value by producing required outputs from a variety of inputs.

Table 2.1 provides a distinction between the different terms used in the context of Process Management and Benchmarking, with *Figure 2.4* providing a pictorial representation.

TABLE 2.1 Three types of measure for process

Terminology	Explanation
1. Process or Efficiency	Resources consumed in the process relative to minimum possible levels that could be obtained
2. Outputs or Effectiveness	Ability of a process to deliver products or services according to specified specifications
3. Outcome or Product/service effectiveness and customer satisfaction	Ability of outputs to satisfy the needs of clients

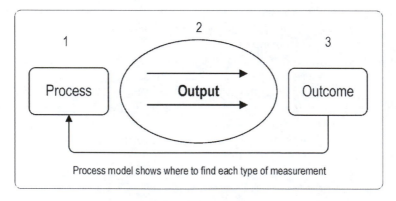

Process model shows where to find each type of measurement

FIGURE 2.4 Process model related to Table 2.1

Benchmarking provides a means for construction organisations to gauge how well processes are performing, relative to both internal and/or external organisations that carry out similar activities.

Benchmarking can also prove very useful to construction firms when they are trying to:

- obtain an objective assessment of their process(es) strengths and weaknesses;
- find ways to stimulate people and groups to engage in improvements;
- overcome internal and external resistance to identified necessary change activities; and
- validate (or not, as the case may be) the methods, operations or resources currently utilised.

Benchmarking also assists construction organisations in answering the 'How do you know?' type of questions, such as:

- How do you know that you are achieving a superior performance?
- How do you know your improvement plan will improve corporate and project performance?
- How do you know that you have the best processes available?

The Public Sector Benchmarking Service identifies the following seven main approaches to benchmarking:

Strategic benchmarking

This is used where construction related organisations are seeking to improve their overall corporate and project performance, by focusing on specific strategies or processes. The key driver for benchmarking is the continuous enhancement of the firm in meeting its strategic aims. And benchmarking is implemented within the context of the development of these core business strategies. Benchmarking is usually undertaken against known examples of best practice, related to the set industrial context.

Performance or competitive benchmarking

This is a process whereby construction firms use performance measures in order to be able to compare their results against those of other similar companies or processes. This is a common practice in most industrial sectors; measures may include cost per unit of production or profit produced per employee. Benchmarking using this methodology can also be applied within a construction organisation by comparing the performance of individual projects or teams.

Process benchmarking

This approach focuses on specific utilised processes or operations, for example, in construction it could relate to the process of materials handling, with a view to determining and deploying improvements.

Functional and generic benchmarking

Functional or generic benchmarking involves partnerships of organisations drawn from different sectors, all of whom have a desire to improve some specific activity or process. The EFQM has encouraged groups of organisations to work together to benchmark approaches to strategic activities such as knowledge management and process management. (See Chapter 4 of this text book for further information.)

External benchmarking

This form of benchmarking can enable the comparison of a construction organisation's functions and key processes; they are compared against good practice in other organisations. The motivator is usually a search for improvement opportunities in business processes.

Internal good practice benchmarking

This is achieved by establishing good practice organisation-wide, usually through the comparison of internal activities or operations. In the context of business planning, this enables the prioritising of specific process improvement projects, allowing results to be compared across business units or projects, in order to identify internal 'best practice', which is then shared.

International benchmarking

Benchmarking can be undertaken internationally as well as nationally.

This is a practice that seeks to identify opportunities for improvement by making comparisons of product/processes or services relating to cost, quality, time, service level and any other key features required by clients.

The relationships between these different types of benchmarking is shown in *Figure 2.5*.

A valid approach to benchmarking would be to select from an appropriate mix of all of the noted methods, and organisational learning is best done when it is carried out within a spirit or partnership and collaboration that enables all parties to learn from each other.

The relationships illustrate that benchmarking can become a central strategy for strategic direction setting and business planning. Benchmarking then becomes a potential improvement strategy for the development of improvement strategies for any key process of the construction organisation.

Benchmarking focuses attention on what is important for the host company. It further provides an approach for creating improved value and quality, but also for improving resource management and productivity. It strikes a balance between stability and renewal and provides a mechanism for studying other organisations to see how they have developed and managed their processes and functions successfully.

Companies that practice benchmarking tend to be more proactive and have a tendency to be externally focused and close to their stakeholders. They are better

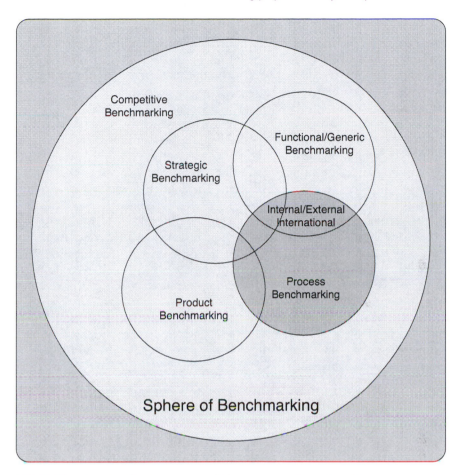

FIGURE 2.5   Relationships between the different types of benchmarking

able to achieve significant improvements in corporate performance and operational competitiveness.

When construction companies embrace the concept of organisational learning, managers and other employees learn and apply new techniques. There is also the opportunity to codify successful behaviour to create new areas of personal competence. Benchmarking can be used as a means to overcome any existing resistance to new ideas, which are often due to staff complacency. Benchmarking can introduce learning as an important component of corporate and project performance. (See Chapter 6 of this text book for further information on organisational learning.)

Inter-firm comparison

Benchmarking can be used to critically analyse any aspect of a construction organisation's project or corporate performance; it is also appropriate for both manufacturing and service industry sectors. Typical organisational functions that are suitable for benchmarking include purchasing, materials control, customer service satisfaction levels and marketing performance and many more.

For around five decades the Centre for Inter-firm Comparison has carried out benchmarking projects and has worked successfully in most sectors of the economy and many countries throughout the world. They are the world leaders in competition benchmarking, sometimes called inter-firm comparison.

Inter-firm comparison is a mutually beneficial activity based on the provision of detailed information, provided in confidence by participating companies, on a comparable basis. Companies decide exactly what information is to be included and exactly how each item is to be defined and identify any differences in company practices that should be taken into account. Therefore the resulting data are valid and can be used with confidence when making comparisons. Typically about one hundred different ratios are utilised and are shown in full for each participating company anonymously, via the application of coded letters. By comparing their ratios with those of other similar companies, it is possible for them to assess their performance against specific set standards. This process can then inform the feedback and feed forward of relevant information leading to continuous project and corporate improvement.

The Centre's inter-firm comparisons are of real practical value to construction firms, at both project and corporate levels. However, firms must ask the correct questions, select the appropriate ratios, critically analyse the results and take timely necessary resulting actions.

Examples of financial benchmarking for construction companies

Performance Ratios

$$\text{Profit Margin } (\%) = \frac{\text{Profit before Tax}}{\text{Sales}} \times 100$$

This ratio is probably the most useful for a construction business, as it enables a firm to see the bottom line for its activities. A construction business will always be interested in how much profit it is making, in comparison to its sales (turnover) and for this reason the profit margin is often expressed as a percentage of the turnover figure. It is not a sensible strategy for a construction company to have a large turnover figure if it is not generating an acceptable profit margin.

$$\text{Return on Shareholder Funds } (\%) = \frac{\text{Net Profit before Tax}}{\text{Total Capital Employed}} \times 100$$

This ratio is obviously important because it establishes for a shareholder or a potential shareholder what he will obtain from his investment in that specific company. Because of the associated risk of investment involved, shareholders will expect a greater return than that which is available from a bank or other financial institution.

$$Credit\ Turnover = \frac{Sales}{Creditors}$$

This ratio provides a means for a construction firm to assess how the firm's turnover relates to how much it owes its creditors. A construction company will normally only want to pay its creditors once it has received payment from its debtors. Thus the figures for Creditor Turnover and Debtor Turnover should stay approximately the same.

$$Debtor\ Turnover = \frac{Sales}{Creditors}$$

See explanation for creditor turnover.

$$Stock\ Turnover = \frac{Turnover}{Stock}$$

This ratio allows the firm to see clearly how the quantity of materials they have in stock compares to their turnover figure. A firm will aim to have as little stock as is practicable, in order to achieve the greatest turnover.

$$Debtor\ Collection\ (days) = \frac{Debtors}{Sales\ Turnover} \times 365$$

This ratio allows a company to calculate how many days on average it is taking to recoup its debts.

$$Creditors\ Payment\ (days) = \frac{Trade\ Creditors}{Purchasers} \times 365$$

This ratio allows a company to calculate how many days it should restrain before paying its creditors, and a useful rule of thumb is that these two ratios should be roughly the same.

$$Net\ Asset\ Turnover = \frac{Sales}{Net\ Sales}$$

This ratio allows a company to assess how much its net assets are worth in relation to its turnover figure. A company does not want to have too much of its value in its fixed assets. If this is the case, it may want to consider other arrangements, such as rental of such assets or the hiring of its plant and equipment.

$$Return\ of\ Equity\ (\%) = \frac{Profit\ after\ Tax}{Shareholders\ Funds\ (capital + reserves)} \times 100$$

This ratio is quite similar to the ratio for return on shareholder funds. It allows the company to assess whether or not it is achieving sufficient profit, based on the amount of money invested by shareholders.

$$\text{Return of Net Assets } (\%) = \frac{\text{Profit before Tax}}{\text{Net Assets}} \times 100$$

This ratio shows how much gross profit a company is achieving in relation to the value of its net assets.

$$\text{Current Ratio} = \frac{\text{Current Assets}}{\text{Current Liabilities}}$$

This ratio is critical because it predicts the solvency of the company. Bankers, auditors, and accountants amongst others will be very interested in the results of this ratio. A construction firm's liabilities should not exceed its assets, as this is not beneficial for the company in the long term.

$$\text{Liquidity Ratio} = \frac{\text{Current Assets less Stocks}}{\text{Current Liabilities}}$$

This ratio is similar to the current ratio, but it provides for an even more stringent test. Banks refer to it as the 'acid test' ratio. It deals mainly with 'liquid' elements, e.g. money. Usually things such as overdrafts would not be included because they are not such an urgent debt, as are creditors.

The results of these financial ratios can thus be compared with similar construction organisations in order to obtain valuable Industrial Comparable qualitative data, to be incorporated into its decision making process.

Financial ratios example

Corporate financial health can be assessed and evaluated by the meaningful comparison of different aspects and components of financial reports, very much upon financial proportions rather than on absolute amounts. There is no single absolute 'dashboard' that indicates 'healthy' or 'poor' ratio performance for an organisation. Differing industries can hold very different views and positions regarding 'acceptable' ratio performance. Whilst a ratio that compares sales (turnover) with value of stock held within the organisation, a stock turnover ratio, might in one industry be considered to be acceptable when the measured relationship is discovered to be 2:1 (the sales value is twice that of the value of stock held), this is not the case for all industries. Industry circumstances vary, for example, whilst supermarkets hold a great deal of stock, the expectation is that that stock is sold many times over in a year. A lean organisation, whatever the industry sector, would seek a high stock to turnover relation. It should be noted that when comparing organisations it is vital that any inter-firm comparison be based on similar types of organisations and businesses from the same business sector.

When interpreting ratios it is important to fully appreciate the composition of the actual items constituting the ratio. For example when considering the solvency of an organisation, the current assets/current liabilities ratio will be considered. It could be that the organisation's solvency is initially perceived as being poor, but with a little further examination and discovery the position is satisfactory as the current assets held are wholly cash balances and not substantially stock or debtors.

It is certainly the case that good financial analysis requires a combination of common sense and sound financial judgement.

Table 2.2 provides an example of financial account data, a condensed profit and loss account and a condensed balance sheet. *Table 2.3* presents an example of the examination of the financial accounts using ratio analysis.

General statement

- The liquidity situation is moving from bad to worse.
- The value of such analysis is improved by comparing and contrasting different financial trading periods to reveal possible trends.

TABLE 2.2 Financial accounts example

Condensed Profit and Loss Account for the year ending 31 December

	Year 2 £000s	Year 1 £000s
Sales	2,400	2,000
Cost of sales	1,400	1,200
Gross margin	1,000	800
Expenses	440	400
Income before taxes	560	400
Less taxation	252	170
Income after taxation	308	230
Dividends	160	160
Income retained	148	70

Condensed Balance Sheet as of 31 December

Capital Employed (Source of Funds)	Year 2 £000s	Year 1 £000s	Fixed Assets Net of depreciation	Year 2 £000s	Year 1 £000s
Share capital – issued and fully paid ordinary £1 share	800	800			
Retained earnings	348	200	Plant/equipment	800	600
9% debenture	100	100	Motor vehicles	300	200
	1,248	1,100		1,100	800
Current liabilities			Current assets		
Creditors	300	270	Inventory	540	400
Taxation	252	170	Debtors	200	180
Dividends Proposed	160	160	Bank	120	320
	1,960	1,700		1,960	1,700

TABLE 2.3 Ratio analysis applied to the financial accounts data

Year 2	Year 1

Profitability Ratios

$$\frac{Profit}{Capital\ Employed} \times 100$$

$$\frac{560}{1,248} \times 100 = 44.87\%$$

An increase of 8.5%

$$\frac{400}{1,100} \times 100 = 36.36\%$$

$$\frac{Profit}{Sales} \times 100$$

$$\frac{500}{2,400} \times 100 = 23.3\%$$

An increase of 3.3%

$$\frac{400}{2,000} \times 100 = 20\%$$

$$\frac{Sales}{Capital\ Employed}$$

$$\frac{2,400}{1,248} = 1.9\ Times$$

An increase of 0.1

Note: Overall profitability shows an increase

$$\frac{2,000}{1,100} = 1.8\ Times$$

Liquidity Ratios

$$\frac{Current\ Assets}{Current\ Liabilities}$$

$$\frac{860}{712} = 1.2:1$$

A fall in long term liquidity of 30p in the £

$$\frac{900}{600} = 1.5:1$$

$$\frac{Quick\ Assets}{Current\ Liabilities}$$

$$\frac{320}{712} = 0.45:1$$

Short term liquidity has fallen by 38p in the £

Note: Quick Assets taken as debtor's + bank cash

$$\frac{500}{600} = 0.83:1$$

$$\frac{Debtors}{Sales} \times 365$$

$$\frac{200}{2,400} \times 365 = 30.40\ days$$

A reduction of 2.45 days

$$\frac{180}{2,000} \times 365 = 32.85\ days$$

$$\frac{Sales}{Stock}$$

$$\frac{2,400}{540} = 4.4\ Times$$

Increase in stock held in relation to sales achieved.

$$\frac{2,000}{400} = 5\ Times$$

Conclusion

This chapter introduced concepts and practices relating to stakeholders, key performance indicators (KPIs) and benchmarking. These all require important consideration when undertaking to measure and improve project and corporate performance. Extended detail and discussion of self-assessment is presented within Chapters 4 and 6 of this text.

Questions for the reader

Here follows some questions related to the information presented within this chapter. Try to attempt each question without reference to the chapter in order to assess how much you have learned. The answers are provided at the end of the text.

Case study question

From the data provided below, Ms Smith, the Managing Director of Smith's PLC, has asked you to explain the current financial situation of her company. She has requested that you use as the basis of your analysis the following *liquidity* and *performance* ratios (KPIs):

Liquidity ratios

$$\frac{\text{Current Assets}}{\text{Current Liabilites}} = \text{Current Ratios}$$

A measure of the level of safety involved in relying on current assets being sufficient to pay current liabilities

$$\frac{\text{Quick Assets}}{\text{Current Liabilites}} = \text{Acid Test}$$

This is a measure of the level of safety involved in relying on 'Quick' assets to cover liabilities.

(Note: 'Quick' assets are usually taken as cash and debtors)

$$\frac{\text{Debtors}}{\text{Sales}} \times 365 = \text{Average time taken to obtain payment from debtors}$$

$$\frac{\text{Sales}}{\text{Stock}} = \text{The number of times stock is turned over per trading period}$$

Performance ratios

$$\frac{\text{Profit}}{\text{Capital Employed}} \times 100 = \text{Return on capital}$$

Here 'profit' is usually taken as being the 'profit before interest and tax' is deducted. This ratio measures the creation of wealth relative to the economic wealth tied up in the process.

$$\frac{\text{Profit}}{\text{Sales}} \times 100 = \text{amount of profit generated by sales made during trading periods}$$

$$\frac{\text{Sales}}{\text{Capital Employed}} \times 100 = \text{the amount of times that the capital is turned over}$$

Ms Smith would like you to support your ratio calculations with a very brief descriptive commentary regarding *Liquidity* and *Profitability*.

Case study: financial accounts

Here follow the summarised financial accounts for the business Monaghan and Monaghan PLC for years ending 31 December 2016 and 2017.

Financial accounts example

TABLE 2.4 Case study financial accounts of Monaghan and Monaghan PLC

Condensed Balance Sheet as of 31 December

Capital Employed (Source of Funds)	2017 £000s	2016 £000s	Fixed Assets (net of depreciation)	2017 000s	2016 £000s
Share capital – issued and fully paid ordinary £1 share	450	450	Freehold & Property (at cost)	336	336
Retained earnings	237	357	Plant/equipment	260	410
11% debenture	300	300			
	987	1107		596	746
Current liabilities			Current assets		
Creditors	641	673	Stock	90	182
Overdraft (secured)	432	266	Work in Progress (jobbing work)	123	132
Taxation	–	34	Uncompleted Contracts (less progress payments)	580	480
			Debtors	671	540
	1073	973		1464	1334
	2060	2080		2060	2080

Condensed Profit and Loss Account for the years ending 31 December

	2017 £000s	2016 £000s
Completed contracts	2,130	2,400
Completed WIP	100	100
Cost of sales	1,875	1,945
Gross margin	355	555
Expenses	375	385
Income before taxes	(20)	170
Less taxation	–	51
Income after taxation	308	230
Dividends	–	–
Net profit/loss	(20)	119

Answers to this case study exercise are provided at the end of the text book.

Further reading

McCabe, S. (2001). *Benchmarking in Construction*. Oxford: Blackwell Science.

McGeorge, D., and Palmer, A.S. (1997). *Construction Management New Directions*. Oxford: Blackwell Science.

Phillips, R.R., and Freeman, E. (2003). *Stakeholder Theory and Organisational Ethics*. Oakland, CA: Berrett-Koehler Publishers.

References

Cook, A. (1999). How do you measure up? *Building*, pp. 24–30.

Force, Construction Industry Task. (1998). Rethinking Construction: Report of the Construction task Force to the Deputy Prime Minister on the Scope for Improving the Quality and Efficiency of UK Construction. London: *DETR*.

Freeman, R.E. (1984). *Strategic Management: A Stakeholder Approach*. Boston: Pitman.

Great Britain, The KPI Working Group (2000). *KPI Report for the Ministers for Construction*. London: Department of the Environment, Transport and the Regions.

McCabe, S. (2001). *Benchmarking in Construction*. Oxford: Blackwell Science.

MacDonald, J., and Turner, S. (1998). *Understanding Benchmarking in a Week*. London: Hodder & Stoughton Educational.

Mitchell, R.K., Agle, B.R., and Wood, D.J. (1997). Toward a theory of stakeholder identification and salience: Defining the principle of who and what really counts. *Academy of Management Review*, 22 (4), pp. 853–886. 10.2307/259247 (http://dx.doi.org/10.2307%2F259247). Available at: www.jstor.org/stable/259247.

Pearson, A. (2002). Don't go KPI nuts. *Building*, 267 (5), 44–45.

3

QUALITY ASSURANCE AND CONSTRUCTION ORGANISATIONS

This chapter introduces quality assurance and advocates that organisations undertake corporate planning to facilitate the adoption of a quality assurance system. It is proposed that competitive advantage can be gained through the application of a quality assurance system. A construction project quality assurance system is briefly outlined and eight quality management principles are identified. Benefits of deploying a quality management system are outlined and problematic issues associated with implementation are identified. Issues concerning an organisation's culture when implementing a quality management system are considered. Finally, a generic model for the implementation of a certified quality assurance system is presented and costs associated with external certification are briefly outlined.

Learning outcomes

By the end of this chapter the reader will be able to demonstrate an understanding of:

- The principles of quality assurance
- Eight quality management principles and the benefits of these principles
- The key issues to be addressed when implementing a quality management system
- The benefits for construction related organisations of deploying a quality management system

Introduction

A strategy for construction organisations seeking to gain a sustainable competitive advantage is that of differentiating products or services. This differentiation can be achieved by the provision of a quality focus. If the providers of products or services are to be competitive, then the quality of their products and/or services must match and surpass customer expectations. This chapter explores how a quality management system can be utilised by construction organisations so as to obtain a sustainable competitive advantage. Benefits and implementation issues associated with quality management systems are identified and a generic process model is presented.

Corporate planning and quality assurance

Most construction organisations engage in some form of corporate planning activity, although some would not recognise it as such. Every small and medium sized company (SMEs) that has decided that it will increase turnover, improve return on investment and increase market share has engaged in the corporate planning activity. In a more sophisticated form, the corporate planning process involves a team of experts working closely with the senior management of an organisation to consider future commercial probabilities and possibilities. A strategy to enable the organisation's future success is developed as a result of this process.

Successful corporate planning is very much dependent upon the provision of accurate and valid information. It requires frequent testing of the feasibility of proposals and constant input from managers at all levels of the construction company.

Corporate plans and human resource (HR) plans must be mutually interdependent and set within a truly holistic context; this has two main facets.

1   Corporate policy shapes HR plans: If corporate policy determines that a company is to expand and diversify, then new employees and skills/competences will probably be required to make this possible. This is an important fact when considering the specific requirements of a quality assurance system.

    Different parts of a business and different functions should be considered separately when examining their HR implications, but a holistic overview must be taken. This will help to prevent the regrettably all too frequent occurrences where one part of a business is declaring redundancies, while in another part a desperate recruitment drive is being undertaken. If such situations are identified in time, corrective action can be taken through re-training and re-deployment to reduce high cost negative consequences.

2   HR plans influencing corporate policy: HR resources can act as a very real constraint upon the achievement of specific corporate objectives and the implementation and continual application of a quality assurance system. Human resource is a valuable resource and as such it must be available in the correct proportion together with the other necessary 4M's (materials, money, machines and management).

When an organisation is assured that it's making the optimum use of its existing human resource and has identified what possible changes in demand for human resource corporate policy may cause (e.g. the deployment of a quality assurance system), it is ready to be proactive. But the identification of human resource needs is not a single once and for all endeavour, it should be a dynamic activity. A whole range of factors can influence the demand for a required number or volume of different types of employees.

These factors include:

- Possible market fluctuations affecting demand for a construction firm's products and/or services and hence the number of employees required to make and/or provide them.
- Changes in the availability of raw materials, affecting levels of production and hence organisational manpower requirements.
- Technological advances which preclude the need for some jobs and possibly change the skills/competences required to perform others. For example a decision to computerise certain systems can usually be considered well in advance of the event, preventing panic reactions. Other changes may require a more immediate response.
- Government intervention (in Health & Safety for example) may lead directly or indirectly to the creation of new jobs or the realignment of some responsibility.
- Mergers and takeovers can affect every aspect of corporate life, and objectives are likely to change, as may the culture of the whole organisation.
- Internal problems such as unexpected industrial relations difficulties.
- Changes in the cost of labour relative to that of other resources.

Each of the above factors forms a variable in an organisation's human resource demand planning and management task. It should be noted that the implementation of a new quality assurance system will have impact upon the organisation's processes and will hence have a corresponding impact upon a construction organisation's human resource requirements.

It is vital that the setting of corporate objectives and their associated impact be given full consideration. However, Johnson and Scholes (1984) have purported: "Corporate objectives are usually formulated by senior members of the board or even the Chairman or Chief Executive. They are more likely to be handed down to lower levels of management [as is the case in most construction companies] than formulated by such lower levels."

Senior management may set the policy and objectives for the company but they should not forget that the implementation issues, especially quality-related issues, involve everyone in the construction organisation. This is a very important point when endeavouring to successfully deploy or refine a quality assurance system.

People within construction organisations are the mechanism by which implementation takes place. The extent to which people within construction organisations influence objectives and deployment has been identified by Miles and Snow (1983), who have postulated: "Organisations [including construction organisations] do not have objectives but people have values." As such, organisational personnel may ultimately have an impact upon corporate objectives. With regard to construction organisational personnel it is necessary to have the correct number of competent staff required to perform set project and corporate tasks. It should be noted though that "[l]abour productivity is a measure of how efficiently the human resources are being used. To some extent it combines an assessment of both efficiency and effectiveness since poor allocation of people to jobs (effectiveness) would result in low productivity" (Johnson and Scholes 1984).

Having an HR staff with appropriate competences and skills will positively impact upon a construction organisation's ability to deliver the required quality specification of products and or services as required by the organisation's customers.

Quality assurance in construction

To be successful, construction organisations need to provide assurance that they have the capability to produce a product that is effective, contractually compliant and commercially economic, whether that product is the *design* of a structure or building and/or the *construction* or *maintenance* and *facilities management* of a structure or building.

The pursuit of quality commences with the client and continues through the production process to the utilisation and management of the building. Quality assurance is an integral part of the "total building process". Quality is in many ways a subjective entity and to a certain extent is a matter of personal judgement. The notion and concept of quality is considered in Chapter 1 of this text. In seeking to provide a clearer view of the meaning of quality in a construction context, Griffith (1990) defines a number of aspects which require consideration, these are: "Function: does the building meet the requirement? Life: is the building durable? Economy: does the building represent value for money? Aesthetics: is the building pleasing in appearance and compatible with its surroundings? Depreciation: is the building an investment?"

Within the context of the construction industry, like other industries, the interpretation and measurement of quality can be as ambiguous as the perceptions and notions of quality. Clients hold and express their own idea of the quality required to meet stated needs and desires.

The architect's aim can be considered to be that of delighting the client whilst delivering value. The build-process in terms of *quality* is very much informed by compliance with regulatory standards as well as conformance with the design and skill in terms of sequence and time with regard to the project programme or schedule. Project delivery on site is subject to the skills and application of the operatives (*workmanship*) and the quality of materials used. When considering the notion of quality in construction, it could therefore be hypothesised that the perception of what quality is and what is important in terms of *quality* is dependent upon an individual's role, involvement and location within the construction process.

Quality in construction can be said to be determined by role and expectation. Dayton (1988) emphasises that "management of quality and quality itself are closely related to a number of various expectations surrounding the performance of buildings, these being quality, durability and reliability".

Quality assurance systems

Quality assurance is concerned with planning and developing the technical and managerial competence to achieve the desired objectives, whether these are set by a client or a construction organisation. Quality assurance is also concerned with the management of people, addressing the roles, duties and responsibilities of individuals within the construction organisation. Quality assurance is primarily the responsibility of management, but its structure and implementation must become part of the total holistic construction organisational framework, and as previously noted, related to the corporate human resource strategy.

Quality assurance must also be an important aspect of the marketing and promotional strategy of the construction firm. Only when quality assurance pervades the entire construction organisation and becomes an integral and recognised aspect of its operations will quality assurance foster the potential to become truly successful in providing an organisational competitive advantage.

Quality assurance must also be actively employed throughout the total building process, from initial feasibility, briefing and conceptual stages, throughout the assembly process, to the completion of the project and the operation of the asset. It is essential that clear communication is planned for and encouraged, in particular at the critical interfaces of project responsibility and control.

Quality assurance application

Quality assurance is concerned with developing a formal structure, organisation and operational procedures designed to ensure the specified quality is attained throughout the total building process. The construction industry can be divided into five broad quality assurance sectors, these being:

- The client, in the production of the project brief;
- The designer, in the design and specification process;
- The manufacturers, in the supply of materials, products and components;
- The contractors (and subcontractors), in construction, supervision and management processes; and
- The user, in the utilisation of the new structure.

There are few standards and codes that affect the client and the final procurement and use of the building, with the majority of quality assurance applications being more related to the manufacturing arm of the construction industry.

Successful implementation of a robust quality assurance system that addresses the five sectors identified above can be certified by a recognised body such as ISO. Such independent certification can serve as competitive advantage when seeking to win work and deliver quality products and services.

Competitive advantage and quality

For a construction organisation to be profitable and sustain growth it must have a sustainable competitive advantage, sometimes referred to as a competitive edge. Johnson and Scholes (1984) identified three valid strategies for organisations to attain a sustainable competitive advantage; these are:

1. Least cost
2. Focus
3. Differentiation

A least cost strategy

A least cost strategy is based upon the reduction of fixed and variable overheads per unit of production or service provision. This enables the manufacturers or service providers to offer their goods or services at a lower price per unit than respective competitors. Least cost is usually associated with high volume production or provision. It is built upon the foundations of *economies of scale*.

Focus as a strategy

Focus, as a sustainable competitive advantage, advocates that an organisation concentrate upon what it is good at, i.e. its distinctive competence. If a construction organisation is good at, for example, small renovation works, it would make little strategic sense to diversify into Civil Engineering. Peters and Waterman (1982) have termed this "Sticking to the Knitting [and note] . . . least successful as a general rule are those companies that diversify". There are certain advantages here related to Learning Curve Theory.

Differentiation as a strategy

Differentiation implies that a construction organisation's product or service is differentiated in some way from its competitors. It is possible that a product or service can be differentiated via a quality aspect. A product or service may be taken as being synonymous with quality (or your customer's definition/perception of quality). Quality assurance certification such as ISO is very concerned with being able to demonstrate that a construction organisation can deliver customer quality requirements.

The above three noted means for obtaining a strategic competitive advantage (least cost, focus and differentiation) are not mutually exclusive. It is possible for a construction company to focus upon a particular product or service by which differentiation is obtained via a quality focus. Here quality assurance certification such as ISO could be part of the competitive strategy. If the service chain becomes large enough and/or a large volume of products are manufactured, then the noted advantages of least cost may also manifest themselves.

Advocacy of quality

The above three strategies highlight that quality issues for both product and service providers are of great importance in seeking to gain competitive advantage. A construction company can reduce its price to increase its market share, but that could prove to be a high risk strategy or tactic. Competitors can quickly follow suit, and not only is the organisation back where it started, but it is likely to be worse off, having started a price war.

Quality is more than the perceived intrinsic quality of a product or service. Quality engages every activity of the construction organisation. The commitment of every person in the construction company is required and this usually leads to a change in the attitudes, behaviours and relationship of the organisation's management and workforce.

If this can be obtained, then quality improvements are feasible and deliverable. The specific benefits of a quality focus have been identified by the Plastics and Rubber Institute:

> Benefits are company wide involvement, improved productivity and efficiency, reduced costs and the ability to provide a product that consistently meets customers' needs, so improving customer confidence and service and retaining or increasing market share.
>
> *(The Institute of Quality Assurance 1990)*

Quality does not, however, just happen, but requires a professional approach and positive action by senior managers, engineers and technologists in every department of a construction company. Dale (2004) notes that "What we are talking about here [truly integrating a quality approach] is a long term cultural change."

Quality assurance and construction project roles

The following provides an overview of the roles undertaken by on site construction staff members in relation to a certified quality assurance system such as ISO. A generic overview is presented within the context of this chapter.

Project manager/site manager

Project managers/site managers carry the burden of accountability for ensuring all site staff under their direction perform their duties in compliance with the project quality assurance system. They are the main link between head office, the client and the project, and as such it is their responsibility to ensure the successful completion of the project, to the complete satisfaction of the client (in accordance with the terms and conditions of the contract documents).

Project managers/site managers ensure that all work carried out conforms to the specified standards and design criteria. They are charged with enforcing the quality assurance system throughout the project, with due regard being given to the performance of subcontractors and ensuring conformity to specified standards by all subcontractors and suppliers.

Any discrepancies in information or flaws in design should be notified to the project administrator for their consideration as part of the quality assurance process. All internal and external correspondence, on site checks and matters arising which might affect the successful completion of the project are to be fully documented and acted upon in accordance with the quality procedures manual.

Any action necessitated out of the quality assurance documentation process must be initiated and progress monitored in the appropriate manner. All quality procedures initiated have to be communicated to the quality manager for inclusion into the company quality assurance system.

Site engineer

The site engineer has responsibility for setting out of the works as directed by the site manager, and must carry out their duties in full compliance with stated specifications and design criteria. Site engineers are required to fully document all their site activities in the required manner, keeping detailed records of:

* Site levels
* Oral instructions from the client's engineer (confirmed in writing at the earliest possibility)
* Discrepancies in information supplied by the client's representative
* Events or findings necessitating a deviation from current designs or specifications reporting any non-conformity directly to the site manager.

Quantity surveyor

The main responsibility of the quantity surveyor is the financial control of the project. When carrying out measurements for payment purposes the QS should note any defective workmanship or non-conformity with the required standards or specifications. The quantity surveyor needs to keep the site manager informed of any such items, discussing possible causes and any remedial action required.

Trade and contract foremen

All trades foremen or subcontractor's foremen have a responsibility to ensure that work carried out by persons under their direction is done to the required standard as specified by the site manager, and with due regard to plans, schedules and specifications, reporting any non-conformity directly to the site manager. All defects or non-conformity identified during the course of their work must be documented in accordance with procedures for submission to the site manager.

Stores foremen

The stores foremen's responsibilities include the checking of all invoices and materials for correctness to specifications, when delivered to site. They should also ensure the correct storage of all materials, reporting defects or non-conformities arising from deviations in procedures directly to the site manager. These should also be fully documented.

Site operatives

Site operatives have a responsibility to perform their works to the required standard and as specified by their foreman or site manager, thus they must have the prerequisite skills and competences (again, a key aspect of an organisation's human resource plan and quality assurance requirements).

During the execution of their duties they are to report defective materials or contradictions of information to their foreman or the site manager.

Outline of a project quality assurance system

The project quality assurance file details the project particulars for use during the effective management of a construction project. The quality assurance system needs to be documented in a way which makes all project information easily accessible to the user, and ensures consistent use of the quality assurance documentation procedures, for all site functions. Company intranets provide an excellent platform for the hosting and sharing of quality assurance documentation.

Quality assurance project file overview

Project details

This section on project details lists the general information required for reference purposes by the site management team; this should consist of the following information:

- Details on the client
- Project address
- Brief description of the project
- Commencement date
- Completion date.

Contract directory

The contract directory lists all persons or organisations who are involved in the project, detailing their involvement and giving their address, telephone and fax numbers. This information, together with the project details, is essential for the effective communication of the management team and the co-ordination of all on site participants/activities.

Site management structure

This is usually presented as an organisational tree or flow chart, indicating all site staff and showing the lines of authority and autonomy, making reference to any individuals or department at the head office that provides a supporting role to the individual site staff.

Subcontractors

This section details all subcontractors on site, providing details of foremen, anticipated labour force and details of any pre-contract undertakings or agreements. For example, the Schedule of Attendances will detail which enabling resources are to be provided by the contractor and those enabling resources which are to be provided by the subcontractor. Details of head office organisation and individuals directly responsible for the subcontractors are also noted.

A subcontractor's programme is included and is incorporated into the main contractor's programme.

Materials specifications

The material specifications section details all material types to be used on the project, noting all details of required specifications, storage and handling details. Additional considerations, such as lengthy order times or details of especially expensive and fragile items, are also noted and highlighted.

Project programme

A detailed project programme will be included, giving full particulars of start and finish dates of all contractor and subcontractor durations, milestones, available float durations, time risk allowances, resources allocations and projected valuations.

Quality policy

The quality policy details the construction company's objectives and commitment to quality and states the standard to which the company's quality system conforms, for example ISO. It is imperative that all staff fully understand this document and that it is implemented, as it will form the basis of the third party audit conducted by the appropriate certification body.

Construction Design and Management regulations (CDM)

The company CDM file must be represented in the Project Quality Assurance File and should include details of the following:

- Method statements
- Plant and equipment identification
- Hazard risk assessment
- Subcontractors risk assessment and method statements
- COSHH and risk assessment records
- Health and Safety Executive (HSE) notification of project
- Company Health and Safety (H&S) policy
- Company insurers
- Details of emergency procedures.

Full documentation of the above will not be contained within the project quality assurance file, but a brief summary of each entry with specific reference to the full documentation of each is required.

Quality management principles

Managing an organisation successfully requires a systematic approach. Success can result from implementing and maintaining a management system which is designed to continually improve performance by addressing the needs of all interested parties. Eight quality management principles have been identified to facilitate the achievement of project and corporate quality objectives. These are:

- **Customer-focused organisation** – organisations depend on their customers and therefore should understand current and future customer needs. They have to meet customer requirements and strive to exceed customer expectations. This approach would result in the following benefits:

Key benefits

- increased revenue and market share obtained through flexible and fast responses to market opportunities;
- increased effectiveness in the use of the organisation's resources to enhance customer satisfaction;
- improved customer loyalty leading to repeat business;
- researching and understanding customer needs and expectations;
- ensuring that the objectives of the construction organisation are linked to customer needs and expectations;
- communicating customer needs and expectations throughout the construction organisation;
- measuring customer satisfaction and acting on the results;
- systematically managing customer relationships; and
- ensuring a balanced approach between satisfying customers and other interested parties (such as owners, employees, suppliers, financiers, local communities and society as a whole).

- **Leadership** – leaders establish unity of purpose, direction and the internal environment of the construction organisation. They create the environment in which people can become fully involved in achieving the organisation's objectives, resulting in:

Key benefits

- people will understand and be motivated towards the construction organisation's goals and objectives;
- activities are evaluated, aligned and implemented in a unified way;
- miscommunication between all levels of an organisation will be minimised, if not eradicated;
- considering the needs of all interested parties including customers, owners, employees, suppliers, financiers, local communities and society as a whole is achieved;
- establishing a clear vision of the organisation's future;
- setting challenging but achievable goals and targets;
- creating and sustaining shared values, fairness and ethical role models at all levels of the organisation;
- establishing trust and eliminating fear;

- providing people with the required resources, training and freedom to act with responsibility and accountability; and
- inspiring, encouraging and recognising people's contributions.

- **Involvement of people** – people at all levels are the essence of a construction organisation, and their full involvement enables their abilities to be used for the organisation's benefit and shall empower the attainment of:

Key benefits

- motivated, committed and involved people within the organisation;
- innovation and creativity in furthering the construction organisation's objectives;
- people being accountable for their own performance;
- people eager to participate in and contribute to continual organisational improvement;
- people understanding the importance of their contribution and role in the construction organisation;
- people identifying constraints to their performance;
- people accepting ownership of problems and their responsibility for solving them;
- people evaluating their performance against personal goals and objectives;
- people actively seeking opportunities to enhance their competence, knowledge and experience;
- people freely sharing knowledge and experience; and
- people openly discussing problems and issues in a drive to learn and improve.

- **Process approach** – a desired result is achieved more efficiently and effectively when related resources and activities are managed as a process, resulting in:

Key benefits

- lower costs and shorter cycle times, through the effective use of resources;
- improved, consistent and predictable results;
- focused and prioritised improvement opportunities;
- systematically defining the activities necessary to obtain a desired result;
- establishing clear responsibility and accountability for managing key activities;
- analysing and measuring of the capability of key activities;
- identifying the interfaces of key activities within and between the functions of the construction organisation;
- focusing on factors such as resources, methods, and materials that will improve key activities of the construction organisation; and
- evaluating risks, consequences and impacts of activities on customers, suppliers and other interested parties and managing accordingly.

- **System approach to management** – identifying, understanding and managing a system of interrelated processes for a given objective, this contributes to the effectiveness and efficiency of the construction organisation, and enables:

Key benefits

- integration and alignment of the processes that will best achieve the desired results;
- results in an ability to focus efforts on the key processes;
- providing confidence to interested parties, as to the consistency, effectiveness and efficiency of the construction organisation;
- structuring a system to achieve the organisation's objectives in the most effective and efficient way;
- understanding the interdependencies between the processes of the system;
- structured approaches that harmonise and integrate processes;
- providing a better understanding of the roles and responsibilities necessary for achieving common objectives and thereby reducing cross-functional barriers;
- understanding organisational capabilities and establishing resource constraints prior to taking action;
- targeting and defining how specific activities within a system should operate; and
- continually improving the system through measurement, evaluation and reflection before taking actions.

- **Continual improvement** – continual improvement should be a permanent objective of all construction organisations, and this should provide:

Key benefits

- a performance advantage through improved organisational capabilities;
- alignment of improvement activities at all levels to an organisation's strategic intent;
- a flexibility to react quickly to opportunities;
- employing a consistent organisation-wide approach to continual improvement of the construction organisation's performance;
- providing people with training in the methods and tools of learning and continual improvement;
- making continual improvement of products, processes and systems an objective for every individual in the construction organisation; and
- establishing goals to guide, and measures to track, continual improvement;
- recognising and acknowledging improvements.

- **Factual approach to decision making** – effective decisions are based on the logical and intuitive analysis of data and information and incorporated into an effective decision making process, thus providing the following:

Key benefits

- informed and implementable decisions;
- an increased ability to demonstrate the effectiveness of past decisions through reference to factual records;

- increased ability to review, challenge and change opinions and decisions;
- ensuring that data and information are sufficiently accurate and reliable;
- making data accessible to all those who need it;
- analysing data and information using valid methodologies; and
- making decisions and taking action based on factual analysis, balanced with experience and intuition.

- **Mutually beneficial supplier relationships** – mutually beneficial relationships between the construction organisation and its suppliers enhance the ability of both organisations to create value, resulting in:

Key benefits

- flexibility and speed of joint responses to changing market or customer needs and expectations;
- optimisation of costs and resources;
- establishing relationships that balance short-term gains with long-term considerations;
- pooling of expertise and resources with partners;
- identifying and selecting key suppliers;
- clear and open communication;
- sharing information and future plans;
- establishing joint development and improvement activities; and
- inspiring, encouraging and recognising improvements and achievements by suppliers.

(Adapted from BSI 1999)

Organisational structure

The structure that is adopted by a construction company must allow for the utilisation of quality procedures and practices. However, managers should take note of the observations of Kanter:

Restructuring the company can entail threats to current productivity such as

1 The cost of confusion – people can't find things, they don't know their own telephone extension . . .
2 Misinformation – communication is haphazard. . . . Rumours are created and take on a life of their own.
3 Emotional leakage – managers are so focused on the tasks to be done and decisions to be made that they neglect or ignore the emotional reactions engendered by change. But the reactions leak out in other ways, sometimes in unusual behaviour.
4 Loss of energy – any change consumes energy – especially if the restructuring is perceived negatively.

5   Loss of key resources – some companies handle consolidation in bureaucratic rather than human ways by establishing uniform policies.

(Kanter 1989)

The above accepted it may still be necessary for some restructuring of the construction
organisation to take place. "The lack of business success [and quality conformance]
in many cases can be attributed to a persistence with an out-dated organisational
structure in almost entirely new circumstances to which the organisational form is no
longer capable of responding both efficiently and effectively" (Pilcher 1986).

The inference from Pilcher is that performance is linked to structure. The performance
stated here is not only the total organisational performance but also that of the individual
manager. This fact is corroborated by Mitzberg (1983). "Just keeping the structure together
in the face of its conflicts also consumes a good deal of the energy of top management."

There may well be some resistance to change within the organisation when
deploying a new quality management system. Bowman and Asch (1989) identify
the following factors influencing perception of change and responses:

A   Change Factors – content and effect of change, speed and method of imple-
mentation.
B   Personal Factors – general attitudes, personality, self-confidence, tolerance of
ambiguity.
C   Group Factors – group norms, group cohesiveness, superior's reaction.
D   Organisational Factors – change history, organisational structure and climate.

Most resistance can be overcome by allaying the employee's fear of change. This can
be done by arranging briefing sessions for all concerned and keeping all informed
as to both the organisational and personal benefits of the deployment of a quality
management system.

When considering what form of structure the organisation might transition to,
a thought from Murdick et al. (1980) should be considered. "Is the form the most
effective from the viewpoint of both strategy and prompt response to competition?"
Furthermore one must not forget that "the organisation must be capable of operating in a dynamic operational environment [especially in construction]. Structures
are no longer viewed as a rigid definition of hierarchical levels" (Kilmann 1985).

There are various types of organisational structure available for selection by
construction organisations. Though many organisations may use a hybrid of the
classical forms, senior managers must not forget that "an organisation cannot function without communication. Communications tie together the component parts
of an organisation" (Chilver 1984).

The structure must be conducive to strategy. "Organisations are full of surprises
because they are so hard to predict" (Bolman and Deal 1984). This is true because
of the human element. "An organisation is a social system deliberately established
to carry out some definite purpose. It consists of a number of people in a pattern of
relationships. . . . Every organisation has a program – a set of planned activities that
can go well or badly. . . . The manager of an organisation is the person who has the
primary responsibility for making its activities go well" (Glassman 1978).

In undertaking responsibility for the purposeful design of an organisation, senior managers must decide how much specialisation and co-ordination is necessary in order to attain the advocated benefits of a quality assurance system and competitive advantage.

Implementing a quality management system – a process approach

Quality management systems require construction organisations to have a quality manual, which details or references the documented procedures. The manual should include a description of the sequence and interaction of the processes that make up the quality management system. It is the construction organisation itself that determines the type and extent of documentation needed to support the operation of the processes that make up the quality management system.

Procedures provide the means of monitoring and controlling the process, and process control will need to be evident at various stages. Juran (1988) writes: "Process control can take place at several stages of progression including set up (start up) control, running control, product control, facilities control".

Process control in a construction service can be somewhat different from that in manufacturing and it is vital that all discrete stages are identified, so as to ensure that no operation is omitted.

Management hierarchy

Formal authority of a manager is when the authority is viewed as originating at the top of a construction organisation's hierarchy and flowing downward through the hierarchy via delegation. Informal authority is the right conferred upon the manager and his subordinates. Butler (1986) notes: "I recognise that hierarchies are essential". However, the real source of authority possessed by an individual lies in the acceptance of its exercise by those who are subject to it. Formal authority is therefore in effect nominal authority (Herbst 1976).

It is a responsibility of senior management to explain to everyone in the construction organisation that the system's approach to quality is not a stick with which management shall beat them. It is a tool which can aid them to achieve success within their operational environments.

Quality policy and objectives

Generally the responsibility for policy formulation rests with the highest level in the construction organisation. However, "There is considerable controversy as to whether policy is basically concerned with setting the goals of an organisation or with establishing a system of rules subject to which goals will be achieved". The policy statement will incorporate objectives and:

> Organisational objectives give direction to the activities of the group and serve as a media by which multiple interests are channelled into joint effort.

Some are ultimate and broad objectives of the firm as a whole; some serve as intermediate goals or sub-goals for the entire organisation; some are specific and relate to short term aims.

(Massie 1989)

The quality policy is the driving force of the system and commits the construction organisation to both meeting stated requirements and improvement. This will become one of the key documents against which the performance of the quality system will be judged. The translation of the quality policy into practice is then facilitated by the definition of supporting objectives. Quality objectives are now a clear requirement in their own right as opposed to a part of quality policy. Management must identify its policy upon the quality issues. They must be established widely within the organisation, support the policy, be measurable and focus on both meeting product requirements and achieving continual improvement.

Senior management may set the policy and objectives for the company but senior management should not forget that the deployment issues involve everyone in the organisation, especially the quality issues. People within construction organisations are the mechanism of implementation.

Quality planning

Quality planning functions at two levels. At the senior management level, it is their responsibility to ensure that the following take place:

- the planning of the quality management system;
- the achievement of continual improvement; and
- the setting of quality objectives.

At a lower level, the organisation's quality documentation in relation to planning for the realisation of quality management processes is mandatory. The format of the quality plan is optional and quality plans only need to be as complex as the product or service demands.

Training and competence

Management should ensure that all employees are trained so that they may perform and implement the stated company objectives. New employees must also be made aware of the company objectives and trained if necessary. The emphasis is clearly on competence rather than just training. The comprehensiveness of training is dependent upon the company and should be embraced within an effective Human Resource Strategy. A construction organisation must establish its training requirements for personnel. After identifying the requirements, it is necessary to plan and implement training programmes. All training achievements should be recorded so that records may be updated and gaps in training and competences established and addressed.

Information and communication

Organisations must ensure effective internal communication between functions regarding system processes and external communication with customers. This applies not only at the contract stage but also with respect to the provision of product information and in the obtaining of feedback. It follows "for information to be useful it has to be accurate, valid and timely" (Bedworth and Bailey 1982).

Customer perception

Customer perceptions must be addressed and this requires sufficient information on satisfaction or dissatisfaction to be gathered. This enables the construction organisation to monitor customer perception on whether or not customer requirements are being satisfied. Having no complaints may only mean that the construction organisation has no information, not that customer satisfaction has been achieved.

Benefits of quality management system deployment

A summary of the advocated advantages to construction organisations of the deployment of a quality management system includes the following:

(i)   Provides a marketing focus for construction enterprises.
(ii)   Provides a means of achieving a top quality performance in all areas/activities of the construction organisation.
(iii) Provides clear and valid operating procedures for all staff.
(iv)   Critical audits are performed allowing for the removal of non-productive activities and the elimination of waste and hence non-value-adding activities.
(v)   Provides a corporate quality advantage acting as a corporate competitive weapon.
(vi)   Develops group/team spirit within the company and thus leads to enhanced staff motivation.
(vii) Improvement of corporate communication systems within an organisation.
(viii) Reduced inspection costs, and hence improved corporate profitability.
(ix)   More efficient and effective utilisation of scarce resources.
(x)   Recognition of Certification, leading to the possibility of obtaining more work.
(xi)   Customer satisfaction, i.e. provide the required customer quality every time and hence attain possible re-engagement.

The above present a strong case for construction organisations to pursue quality assurance certification. It may also be the case that a prospective client has a requirement that whoever they are going to do business with must have certification. A company can be excluded from a tender list for not having a certified quality assurance system. This in itself is a strong rationale for seeking certification.

Problematic issues associated with the implementation of a quality management system

This section of the chapter focuses on the practical problematic issues associated with implementing a quality management system.

The previous section has established the main advocated advantages of implementation. However, the implementation process can prove to be a most problematic one, and the following establishes the critical issues and provides advocated solutions designed to assist construction related organisations in the implementation process.

It cannot be overstated that the support of senior management is a vital element required for the successful implementation of a quality management system. If senior management support is not forthcoming, the quality facilitator/manager (the person charged with the implementation of the quality management system) could also face further problematic issues, such as:

- A lack of adequate authority to get people fully engaged.
- Insufficient funding for the project, leading to inadequate resources being allocated.
- Lack of sufficient time being allocated for the project; this is a vital resource and has to be provided for.
- Resistance to
  (i) information gathering and documentation production stages;
  (ii) the implementation process during the project.

Total commitment from senior managers needs to be demonstrated and championed through 'policies' and 'overt support'.

If construction organisations are to avoid problems relating to resource issues, senior management must provide the necessary resources. Senior management must take an active role in both designing and implementing the quality system, with support coming from the very start of the project.

The two most important resource issues are adequate funding for the project and the allowance of sufficient time for people to participate. Participation is necessary when the quality facilitator is gathering information for writing of the quality and procedures manuals. Participation of staff is also vital during the implementation phase of the project. It should be noted that time allocation and funding are not mutually exclusive. A lack of funds can mean that money is not available to release staff when participation is required. Issues of authority and overcoming resistance to change are also not mutually exclusive. "If appropriate authority does not accompany managerial responsibilities and duties, the manager's effectiveness within the organisation is impaired" (Glassman 1978).

Glassman purports that managers should be delegated sufficient authority to complete their allocated tasks. Senior management need to ensure middle managers are not asked to perform tasks for which they have not been given the necessary authority to complete. There may well be some resistance to change within the host organisation. Coalitions of resistance could develop and if they are linked to a power base they could impede the implementation process.

The quality facilitator should try and overcome resistance by allaying employees' fear of change.

Managers within the construction organisation must 'manage'. They should not abdicate the responsibility to the quality facilitator (team) without providing adequate authority. Liebmann offers support here. When he was part of a quality team implementing a quality system, he found that senior managers were:

> [c]harged as to design a process to empower employees but did not empower the team. The result was failure of the 'Quality Project.'
>
> *(Liebmann 1993)*

Even before the implementation process begins, staff need to be made aware of the benefits of certification. They need to be convinced that the introduction of a quality system is worthwhile and can provide advantages for them and the organisation. It is, therefore, senior management's duty to echo the rationale for the advantages of certification. This is an important issue, since "people tend to have an in-built resistance to change".

The co-operation of staff is vital for successful implementation and in order for them to co-operate two issues require attention and consideration:

- staff have to want to co-operate; and
- staff have to be allowed to co-operate.

If staff are not coerced into co-operation, they will make a greater contribution to the implementation process.

It can be concluded here that senior management support is a vital component at all stages of the design and implementation of a quality management system. If this support is not provided, a successful outcome will not be attained.

Issues with an organisation's culture

Quality systems require an organisation's culture to possess trust and a desire to identify and eliminate problems. The concept of empowerment is a key component of an effective quality culture of an organisation.

If a climate of distrust exists between senior management and the rest of the organisation, then implementation of the quality assurance system is doomed to fail. The organisation's culture informs the way a business operates, and how employees respond and are treated. Organisational culture contains such contributory components as guiding philosophy, core values, purpose and operational beliefs. It must be understood that just following documented procedures and complying with standards will not guarantee success. Only if the correct culture exists will the true benefits be attained. A culture based upon morphogenic principles is required.

Whoever is charged with the task of designing and implementing a quality system must have the total support of the organisation. This support involves not only senior management but also the employees (the people who perform the documented tasks). If those responsible for implementation can obtain this total support for the system, then successful implementation is possible. An important part of obtaining total involvement is to inform people of what the system is all about and to keep them informed throughout the design and implementation process. Successful implementation of any quality model depends upon the co-operation of all the people who are involved with it.

The following is an overall generic strategy for quality assurance system implementation, and is of value to construction organisations embarking upon the deployment process:

- obtain support from the total organisation/stakeholders;
- set realistic objectives/quality checks, such as timescale for the implementation process;
- deployment is a project, so plan and programme activities ahead of time (engage project management principles);
- maintain internal and external contacts with key personnel/stakeholders;
- establish a clear review/monitoring feedback process;
- be flexible and willing to sacrifice time and other resources to obtain improvements;
- do not expect a great improvement in the saving of resources immediately. Have realistic targets;
- use expert opinions and advice when necessary; you may have to engage external consultants; and
- do not expect too great an immediate return on investment. Some improvement projects may have key benefits because they provide customer satisfaction and assist the long-term survival of the organisation.

Beck and Hillmar (1986) provide a worthwhile note: "A manager needs to be clear with employees about their roles and responsibilities and the results expected in order for them to know what they are accountable for. While holding employees accountable for performance the manager must be accountable to them for support".

Figure 3.1 provides a generic model for construction companies wishing to deploy a quality assurance system and gain the advantages of external certification.

Costs associated with quality management certification

Potential benefits from establishing and maintaining a certified quality management system are not secured without costs to the construction organisation.

These costs are both direct and indirect and include:

Direct

- Developing the quality management system.
- Producing the quality documentation.
- Establishing the implementation system.
- Maintaining the internal audit system.
- Independent third-party assessment.

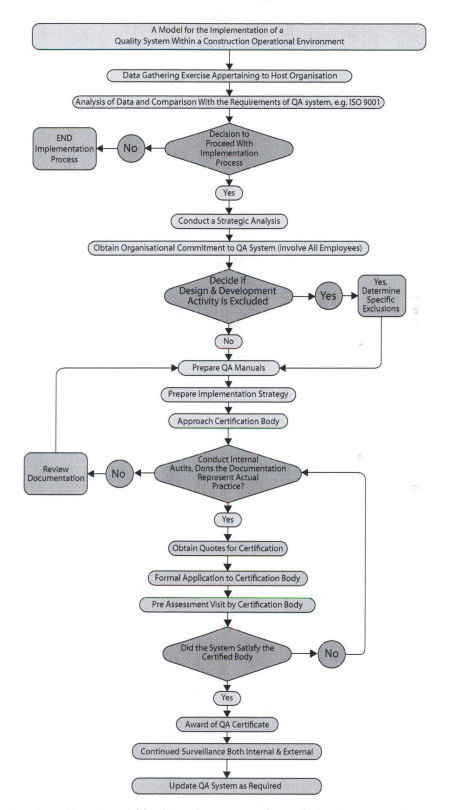

Figure 3.1 Generic model for the implementation of a certified quality assurance system

Indirect costs are difficult to assess but can include

- Liaising with the certification body.
- Changes to operational processes and procedures to accommodate certification requirements.
- Some demotivation aspects associated with staff and the implementation process.
- The consumption of organisational energy and efforts during the drive for certification.
- There are also the costs of maintaining the system and surveillance visits by the certification body.

Certification bodies specify their various registration fees, which are subject to some variation depending on the following factors:

- The size of the company and number of employees.
- The structure of the organisation.
- Diversity and range of the company's activities.
- Nature and complexity of the quality system.
- Complexity of the documentation.

Conclusions

Quality management systems can help construction organisations to obtain a competitive advantage. A quality management system can enable the company to operate more efficiently and effectively. This chapter has considered how a quality management system can be utilised by construction organisations to obtain sustainable competitive advantage. Furthermore it has presented quality assurance and associated key benefits and implementation issues. The importance of an organisation's structure has been discussed and finally the chapter has presented a generic process model for construction organisations wishing to deploy a quality assurance system and gain advantages of external certification.

Questions for the reader

Here follows some questions related to the information presented within this chapter. Try to attempt each question without reference to the chapter in order to assess how much you have learned. The answers are provided at the end of the text.

Question 1

The Construction Industry can be divided into five broad sectors where quality assurance is applicable; identify these sectors.

Question 2

By implementing a certified quality management system, a construction organisation can demonstrate that it has considered and deployed suitable strategies for addressing eight key quality management principles. What are these eight key quality management principles?

Question 3 – case study

The senior management of a construction company has been considering the deployment of an externally certified quality assurance system, as a means of potentially being included on more client tender lists.

The managing director appreciates the value of *quality* as a potential competitive advantage and is a quality advocate within the organisation.

If the implementation of an externally certified quality assurance system is to be successful, why is it essential to have senior management support for the deployment process, and what are the likely outcomes if this support is not forthcoming?

Further reading

Donnelly, M. (1999). Making the difference: Quality strategy in the public sector. *Managing Service Quality*, 9 (1), pp. 47–52.

Hellard, R.B. (1993). *Total Quality in Construction Projects*. London: Thomas Telford.

Hutchin, T. (2001). *Unconstrained Organisations: Managing Sustainable Change*. London: Thomas Telford Ltd.

Naoum, S. (2001). *People and Organisational Management in Construction*. London: Thomas Telford.

Oakland, S. J. (1993). *Total Quality Management*. London: Butterworth Heineman.

References

Beck, A.C., and Hillmar, E.D. (1986). *Positive Management Practices*. London: Jossey Bass Ltd.

Bedworth, D.D., and Bailey, J.E. (1982). *Integrated Production Control Systems*. Chichester: J Wiley.

Bolman, L.G., and Deal, T.E. (1984). *Modern Approaches to Understanding and Managing Organisations*. London: Jossey-Bass Inc.

Bowman, C., and Asch, D. (1989). *Strategic Management*. London: MacMillan Education Ltd.

British Standards Institution (1999). *Quality Management Systems: Fundamentals and Vocabulary*. London. ISO CD 2 9000, Draft.

Butler, G.V. (1986). *Organisation and Management*. Hemel Hempstead: Prentice Hall UK Ltd.

Chilver, J.W. (1984). *People, Communication and Organisation*. Oxford: Pergamon Press Ltd.

Dale, B.G. (2004). *Managing Quality*. 4th ed. Oxford, UK: Blackwell Publishing.

Dayton, J.B. (1988) *Quest for Quality: Developments in the management of Quality by the United Kingdom*. Department of the Environment. Property Services VOL. 1. Agency. CIB W/65, Organisation & Management of Construction Proceedings. Department of the Environment.

Glassman, A.M. (1978). *The Challenge of Management*. Toronto: J. Wiley and Sons.

Griffith, A. (1990). *Quality Assurance in Building*. London: Macmillan Education Ltd.

Herbst, P.G. (1976). *Alternatives to Hierarchies*. Netherlands: Mennen Asten.

The Institute of Quality Assurance (1990). *Training for Quality*. London: The Institute of Quality Assurance.

Johnson, G., and Scholes. K. (1984). *Exploring Corporate Strategy*. London: Prentice Hall International.

Juran, J.M. (1988). *Juran on Planning For Quality.* London: Macmillan Free Press.

Kanter, R.M. (1989). *When Giants Learn to Dance.* Reading: Cox Andwyman.

Kilmann, R.H. (1985). *Beyond the Quick Fix.* London: Jossey Bass Inc.

Liebmann, J.D. (1993). A Quality Initiative Postponed. *New Directions For Institutional Research,* (78), pp. 117–121.

Massie, J.L. (1989). *Essentials of Management.* Englewood: Prentice Hall International.

Miles, R., and Snow, C. (1983). Some tests of effectiveness and functional attributes of miles and snow's strategic types. *Academy of Management Journal,* 26 (1), pp. 5–26.

Mitzberg, H. (1983). *Structure in Fives: Designing Effective Organisations.* Englewood Cliffs: Prentice Hall.

Murdick, R.G., Eckhouse, Moor, R.C., and Zimmerer, T.W. (1980). *Business Policy, A Framework for Analysis.* Indianola: Grid Publishing Inc.

Peters, T.J., and Waterman, R.H. (1982). *In Serach of Excellence.* Cambridge: Harper and Row.

Pilcher, R. (1986). *Principles of Construction Management.* Maidenhead: McGraw Hill.

4

THE EUROPEAN FOUNDATION FOR QUALITY MANAGEMENT EXCELLENCE MODEL

This chapter establishes the linkages between Total Quality Management (TQM) and the European Foundation for Quality Excellence Model (EFQM.E.M). The EFQM Excellence Model is outlined and issues relating to its deployment, self–assessment methodologies and advocated benefits are discussed. The chapter also provides a flow diagram designed by the book's authors, which should prove a useful tool for construction related organisations engaged in the implementation process of the EFQM.E.M.

Learning outcomes

By the end of this chapter the reader will be able to demonstrate an understanding of:

- The linkages between TQM and EFQM.E.M
- The EFQM.E.M and its constituent parts
- The benefits of EFQM.E.M deployment
- The application of the self-assessment methodologies and their part in the attainment of organisational improvement.

Introduction

This chapter explores the rationale for construction organisations engaging in the application of Total Quality Management (TQM). The European Foundation for Quality Management Excellence Model (EFQM.E.M) is introduced and offered as a means of implementing TQM. The EFQM.E.M is linked to self-assessment and continuous improvement.

Total Quality Management

TQM should assist in making the effective use of all organisational resources, by developing a culture of continuous improvement. This empowers senior management to maximise their value-added activities and minimise efforts/organisational energy expended on non-value-adding activities.

TQM enables companies to fully identify the extent of their operational activities and focus them on customer satisfaction. Part of this service focus is the provision of a significant reduction in costs, through the elimination of poor quality in the overall process. This empowers companies to attain a truly sustainable competitive advantage. TQM provides a holistic framework for the operational activities of enterprises. If a firm can overcome problematic implementation issues, then a sustained competitive advantage is the reward to be gained (Watson and Chileshe 2001).

Some TQM proponents maintain that a common error in the application of TQM is the failure to recognise that every company, and environment, is different (Laza and Wheaton 1990 cited by Spencer 1994). Thus successful deployment is dependent upon the correct alignment of corporate strategies and operational environments both encapsulated within a morphogenic culture.

Furthermore, "an important component of TQM is the implementation of work practices such as employee training, information sharing, involvement and empowerment" (Hendricks and Singhal 2001).

TQM requires an organisational corporate culture where change and innovation are expected.

The application of TQM has been advocated by various eminent authors, for example Oakland (1993), Wright (1997) and Cherkasky (1992). However, the process of deployment can prove to be most problematic for many public and private sector organisations. Terms such as empowerment, cultural dynamics and cross functional communications have only served to add to the confusion. Further, many Western organisations have endeavoured to adopt TQM based upon an eastern philosophy and culture and this has complicated matters further. Therefore, a practical application for TQM deployment was required.

The EFQM.E.M is a model for TQM deployment based upon practical application and feedback from practitioners; it has been developed to be most suited for Western business organisations.

In pursuit of best practice

When a construction organisation is seeking to be the best, it must address and answer the following key questions:

Is the construction company striving for

- Better products/services?
- Better organisation/management?
- Better information/communication systems?

These are not mutually exclusive and indeed, a construction company may well strive for all three.

Harrison (1993) outlines the way forward as first creating a vision for the company, second installing ownership of the issues facing the company, and third, planning and implementation of a continuous change process. However, above all, construction companies must "focus on their customers".

Some of the techniques and issues to be addressed by organisations in order to function at best practice level include the following:

The first issue to be considered is that of senior management commitment.

There is no substitute for effective leadership by senior management. The number of failures on the road to implementing TQM and operating at best practice level blamed on a lack of management commitment suggests not only that it is needed but that it cannot be assumed and may be difficult to obtain in practice (Nunney 1992). The important issue of senior management commitment is explored somewhat in Chapter 3 of this text. It is vital that "[p]eople at all levels in the [construction] organisation must fully understand the organisational objectives and the timeliness of the objectives" (Mundy 1992). This is a function of senior management activities. Further considerations for an organisation pursuing a strategy of best practice are the 4P's: purpose, planning, process, and performance measurements.

Purpose

People perform better when they understand the objectives of the company and appreciate that teamwork is an essential element, therefore people should be trained to work in teams. Organisations must embrace the concept of 'empowerment'. People need timely feedback upon their performance, as this affects the intrinsic motivation of individuals.

Planning

Planning and monitoring requires the setting up of a dynamic closed feedback loop. A further requirement is that the Senior Management Team (SMT) must not just think in terms of reducing labour costs per service facility or product manufactured. Consideration should be given to reducing the time taken for materials to undergo each production process. This of course requires planning, incorporated within which should be a collaborative partnership approach towards suppliers, with the attainment of long-term strategic benefits being the main objective.

Process

The process(es) must be flexible and this may demand as much organisational effort in support facilities as it does in the operational activities. Design activities must be incorporated with the production function; this requires a holistic approach.

The process(es) will in most instances determine the service/product quality and such processes must add value. It should be appreciated by all concerned that quality is ultimately everyone's business.

Performance measurements

These require simple dynamic systems. However, it is not the quantity of data that is important but its usefulness in the management decision-making process. In summary it may be stated that the problematic issues are not the technological-based ones but the 'people issues'. "Management acts to develop its people by caring for and training them" (Hickman and Silva 1989).

External and internal changes required of construction organisations

The concept of continuous improvement is an important aspect of a best practice performer and a fundamental function of TQM/EFQM E.M (Yip 1992). Yip notes the dynamic nature of best practice operations. In order for a construction organisation to become and remain a world class performer it must consider both internal and external requirements.

Competition

> The trend towards economic liberalisation in general and privatisation in particular has had a major impact on business activity.
>
> *(Preston 1993)*

In accordance with Preston's statement above, the volume of competition for most construction organisations has increased. As such, construction organisations have a heightened need to understand the nature of their competitive environment. Organisations must operate an all-embracing macro business strategy and not a micro one. Environmental scanning would provide the means for analysing a company's competitive environment.

Suppliers

The way a construction organisation deals with its suppliers must be as a joint venture(s), i.e. in partnership as a collaborative supply chain. Construction companies and their suppliers must understand the synergistic advantages that are available to both parties. A win-win outcome for both parties must be an inter-organisational shared relationship goal.

Environmental factors

The environmental factors must be analysed to the extent that they provide either opportunities or threats to the construction company. The requirements of the environment may involve some physical alteration to the company. A Political, Economic, Social, Technological, Environmental and Legal (PESTEL) analysis would prove useful for organisations.

Economic factors

Construction companies are impacted by changes in monetary value and public expenditure. Such factors can affect profitability and the order book and impinge upon the supply and demand for the services or products.

Technological factors

Considerable interest has been shown in the implementation of new technologies. This is because these technologies are not just for internal consumption but also enable construction organisations to communicate and interact on an international basis in real time. BIM is an example of a technological factor that is having a significant impact upon the construction industry in many dimensions, quality management being one such dimension. Chapters 7 and 8 provide further insight into BIM and quality management.

The internal changes that may be required can be viewed, to some extent, as changing production systems from the traditional 'push system' to a more modern 'pull system'.

Some of the issues raised under internal and external changes do overlap each other. However, one must not forget that the main internal change required is people orientated. "Until recently, most senior operations managers did not perceive the human organisation as a source of competitive advantage" (Ross 1991).

Employees have an in-built resistance to change (Kanter 1989). Nevertheless, change may be necessary. For example, the type of organisational structure the company utilises can affect its ability to function at a best practice level. In many cases, the lack of business success in the deployment of TQM/EFQM.E.M can be attributed to a continued persistence with an outdated organisation structure.

If change is necessary, then "the key to making the transition work is in the employee's understanding of its necessity [and value to them, as well as the company]" (Ross 1991).

Usually groups of people in construction organisations recognise that work could be done more efficiently and/or effectively, but they are rarely asked for their opinion. It is worth noting the thoughts of Ross (1991): "To participate effectively in the global market place, the implementation of production technologies must be combined with a programme aimed at aligning organisational structure and culture, the role and flow of information and people resources, if enterprises [including construction] are to exploit opportunities".

The EFQM.E.M model addresses the points identified above and provides effective linkages between people, agile processes and results (in terms of both past results and confidence in future performance).

The European Foundation for Quality Management Excellence Model (EFQM.E.M)

The EFQM Excellence Model was formerly known as the Business Excellence Model. It was developed between 1989 and 1991 by practitioners to bring together the various models for Total Quality Management (TQM) deployment.

European organisations were experiencing difficulties in the implementation of TQM principles and hence attaining the following benefits of TQM application in practice:

* the production of a higher quality product/service through the systematic consideration of clients' requirements;
* a reduction in the overall process/time and costs via the minimisation of potential causes of errors and corrective actions;
* increased efficiency and effectiveness of all personnel with activities focused on customer satisfaction; and
* improvement in information flow between all participants through team building and proactive management strategies.

The Excellence Model was designed to be:

* Simple [easy to understand and use]; holistic [in covering all aspects of an organisation's activities and results, yet not being unduly prescriptive]; dynamic [in providing a live management tool which supports improvement and looks to the future]; flexible [being readily applicable to different types of organisations and to units within those organisations]; innovative.

(European Foundation for Quality Management 1999b)

The EFQM Excellence Model has been used extensively and beneficially in manufacturing, construction, banking and finance, education, management and consultancy. Companies apply the EFQM Excellence Model, as the pursuit of business excellence through TQM is a decisive factor in enabling competitiveness in today's global market place.

EFQM is a non-profit making organisation providing various networking, benchmarking and training events to help members keep up with the latest trends in business management and research in TQM. It launched the European Quality Award in 1991 to stimulate interest and it is awarded to those who have given 'exceptional attention' to TQM.

The British Quality Foundation is the UK sponsor of the EFQM Excellence Model. The aim of the British Quality Foundation is to promote continuous improvement and organisational excellence using the EFQM Excellence Model. The philosophy of the Foundation is succinctly expressed in the following quote:

Regardless of sector, size, structure or maturity, to be successful, organisations need to establish an appropriate management system. The EFQM Excellence

Model is a practical tool to help [construction] organisations do this by measuring where they are on the path to Excellence; helping them understand the gaps; and then stimulating solutions. The EFQM is committed to researching and updating the Model with the inputs of tested good practice from thousands of organisations [including construction organisations] both within and outside of Europe. In this way we ensure the model remains dynamic and in line with current management thinking.

(European Foundation for Quality Management 2000a)

The model is officially referred to as the European Foundation for Quality Management (EFQM) Excellence Model and has evolved since its introduction after widespread consultation with EFQM members.

The EFQM Excellence Model provides a framework for self-assessment. Using this tool, a construction organisation or indeed any organisation can assess whether it is doing the right things and obtaining the required results. The ensuing assessment of an organisation's performance is measured both by results and the quality of the processes and systems developed to achieve them. In its most sophisticated form, the model is used to assess an organisation for quality awards – including the European Quality Award. The assessment encompasses the whole organisation (or the whole of a part of an organisation) using nine standard criteria. The model provides a balance and a relationship between approach (*enablers* – the ways in which results are achieved) and results (what is achieved in terms of customers, people, society and the business). This provides for a balanced view between cause and effect. The criteria which deal with causes are called *enablers*. Those which deal with effects are known as *results*. In scoring the organisation, enablers and results have an equal 50/50 weighting.

EFQM.E.M

The EFQM,E.M may be viewed in four ways:

- as a framework which organisations can use to help them develop their vision and goals for the future, in a tangible and measurable way;
- as a framework which organisations can use to help them identify and understand the systemic nature of their business, the key linkages and cause and effect relationships;
- as the basis for the European Quality Award, a process which allows Europe to recognise its most successful organisations and promote them as role models of excellence from which others can learn; and
- as a diagnostic tool for considering the health, agility and future of the organisation.

Through use of the model, construction organisations are better able to balance priorities, allocate resources and generate realistic business plans (European Foundation for Quality Management 2000a).

The model can be used for a number of activities, including for example, self-assessment, third party assessment, benchmarking and as the basis for applying for the European Quality Award.

The constituent parts of the EFQM Excellence Model

The model is made up of three integrated components, these being:

- The fundamental concepts of excellence
- The criteria
- The RADAR.

The model is regularly reviewed in a three-year cycle in order to ensure that changes to the model are incremental and that any changes are able to be adopted and implemented with ease by organisations.

An EFQM Management Document is required to support the successful deployment of the EFQM,E.M. The document should be between 20 and 35 pages in length and comprise the following 3 sections:

- Key Information: a summary of the organisation's structure, operating environment, stakeholders and strategic objectives;
- Enabler section: key approaches adopted by the organisation to achieve the strategic objectives; and
- Results section: key results achieved by the organisation, illustrating the effectiveness of progress towards the strategic goals.

The EFQM Excellence Model is based and supported by specific concepts which are referred to as "The Fundamental Principles of Excellence"; these are:

1 Adding value for Customers
2 Creating a sustainable future
3 Developing organisational capability
4 Harnessing creativity and innovation
5 Leading with vision, inspiration and integrity
6 Managing with agility
7 Succeeding through the talent of people
8 Sustaining outstanding results.

Each concept plays a role directly and is indirectly related to different criteria and sub-criteria within the EFQM Excellence Model. There are eight fundamental principles of excellence which inform the nine criteria which an organisation measures its performance against:

1 Leadership (10%)
2 People (10%)
3 Strategy (10%)

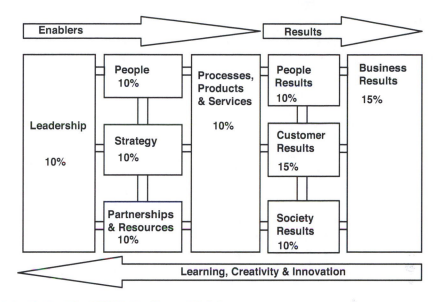

FIGURE 4.1   The EFQM Excellence Model

4   Partnerships and resources (10%)
5   Processes, products and services (10%)
6   People results (10%)
7   Customer results (15%)
8   Society results (10%)
9   Business results (15%).

The model as depicted in *Figure 4.1* consists of two main parts, "**Enablers**" which are made up of five criteria, and "**Results**" which consists of four criteria. Basically, the "Enablers" deal with what an organisation does, while the "Results" deal with what an organisation achieves. Most importantly is that mechanism of feedback within the model in which "Results" are generated by "Enablers" and "Enablers" improved by using the feedback from "Results" (EFQM 1999b).

Model contents and structure

> The EFQM Excellence Model offers an operational tool for the pursuit of excellence in performance and results.
>
> *(European Foundation for Quality Management 1999b)*

The Excellence Model recognises that there are many approaches to achieving excellence (outstanding levels of performance) and provides firms with a way of achieving a top quality performance. The model's nine boxes in *Figure 4.1* represent the criteria against which to assess an organisation's progress towards excellence. Each of the nine criteria has a definition, which explains the high level meaning of that specific criterion.

Five 'Enablers' criteria and four 'Results' criteria, together with their sub-criteria, provide a guide and focus for organisations, so as to succeed in satisfying their respective clients' requirements and achieve outstanding levels of performance.

The nine criteria and sub-criteria

Leadership definition

Leaders shape the future. They are role models for the organisation's values and ethics and enable the organisation's ongoing success.

Five criterion parts support the Leadership criterion:

1a   Leaders act as role models and undertake to develop the mission, values and ethics and act as role models.
1b   Leaders are responsible for the management system and its performance.
1c   Leaders engage with external stakeholders.
1d   Leaders promote a culture of excellence.
1e   Leaders ensure change is well managed and that the organisation is flexible.

Strategy definition

How the organisation implements its mission and vision via a clear stakeholder focused strategy, supported by relevant policies, plans, objectives, targets and processes.

Four criterion parts support the strategy criterion:

2a   Strategy is based on the needs and future needs and expectations of stakeholders and the environment external to the organisation.
2b   Strategy is built on information from internal performance and capabilities.
2c   Strategy is developed, reviewed and updated.
2d   Strategy and policies are developed reviewed and implemented.

One must not forget that implementation is a process undertaken by people and therefore they have to be involved from the start of the process.

People definition

This criterion concerns how the organisation manages, develops and releases the knowledge and full potential of its people at an individual, team-based and organisation-wide level. People capabilities are developed. Fairness, equality, reward and recognition are motivators.

Five criterion parts support the People criterion:

3a   People resources are planned and managed with reference to the organisation's strategy.

3b People's knowledge and competencies are identified and developed.
3c People are involved and empowered at all levels of the organisation.
3d People within the organisation have an effective dialogue.
3e People are rewarded, recognised and cared for.

Partnerships and resources definition

How the organisation plans and manages its external partnerships and internal resources in order to support its policy and strategy and the effective and efficient management of processes and impact upon society and the environment.

Five criterion parts support the partnerships and resources criterion:

4a Sustainable benefit informs the management approach to suppliers and partners.
4b Sustained success underpins the management of finances.
4c Buildings, equipment and materials are managed in a sustainable manner.
4d Strategy is supported by technology.
4e Capability and decision making is informed by information and knowledge.

Processes, products and services definition

These are designed to increase value for customers and stakeholders.

Five criterion parts support the processes, products and services criterion:

5a Processes are systematically designed and managed to maximise stakeholder value.
5b Products and services create maximised value for customers.
5c Products and services are promoted and marketed effectively.
5d Products and services are produced, delivered and serviced.
5e Customer relationships are managed and enhanced.

Customer results definition

What the organisation is achieving in relation to the needs and expectations of customers.

Two criterion parts support the customer results criterion:

6a Perceptions – customer perceptions of the organisation
6b Performance indicators – internal measures to improve the organisation's performance as well as measures to predict the impact of customer perceptions

People results definition

What the organisation is achieving in relation to its people – is it meeting and exceeding needs and expectations?

Two criterion parts support the people results criterion:

7a   Perceptions – peoples' perceptions of the organisation
7b   Performance indicators – internal measures to improve the performance of people predict impact on perceptions

Society results definition

What the organisation is achieving in terms of the needs and expectations of society stakeholders.
Two criterion parts support the society results criterion:

8a   Perceptions – society's perceptions of the organisation
8b   Performance indicators – internal measures to improve the performance of the organisation predict impact on perceptions of relevant stakeholders

Business results definition

What the organisation is achieving in relation to the needs and expectations of business stakeholders.
Two criterion parts support the business results criterion:

9a   Business outcomes – key financial and non-financial measures. Measures and targets are agreed with business stakeholders
9b   Performance indicators – internal measures to improve operational performance

RADAR

At the heart of the EFQM Excellence Model a specific logic exists. This is known as RADAR. The RADAR logic consists of the following four elements: **R**esults, **A**pproach, **D**eployment, **A**ssessment and **R**eview.

The key concepts built into RADAR are that the learner experience is critical to self-assessment, with an emphasis on collecting valid and reliable evidence, and that the primary purpose of self-assessment and development action plans is self-improvement. All assessments are required to be deployed with rigour, irrespective of organisational size. The host organisation should always try and triangulate its collected evidence base, reflection upon results is vital in order to engage in triple-loop-learning.

The application of the RADAR philosophy will assist in driving business improvement through utilising the model. The logic purports that a construction organisation needs to:

TABLE 4.1 Succinct overview of the RADAR concept

Results

This covers what an organisation achieves. In an excellent construction organisation, the results will show positive trends and/or sustained good performance, which will compare well with others and will have been caused by the adopted approaches. Additionally, the scope of the results will address the relevant areas.

Approach

This covers what a construction organisation plans to do and the reasons for it. In an excellent construction organisation the approach will be sound – having a clear rationale, well-defined, developed and integrated processes and a clear focus on stakeholder needs – supporting policy and strategy and linked to other approaches where appropriate.

Deployment

This covers what a construction organisation does in order to deploy the approaches. In an excellent construction organisation, the approaches will be implemented in relevant areas, in an appropriate systematic way.

Assessment and Review

This covers what a construction organisation does to assess and review both the approaches and the deployment of the adopted approaches. In an excellent construction organisation, the approach, and the deployment of it, will be subject to regular measurement, learning activities will be undertaken and the output from both will be used to identify, prioritise, plan and implement improvement activities.

- Determine the **Results** it is aiming for as part of its policy and strategy making process. These results cover the performance of the organisation, both financially and operationally, and the perceptions of its stakeholders.
- Plan and develop an integrated set of sound **Approaches** to deliver the required results both now and in the future.
- **Deploy** the approaches in a systematic way to ensure full implementation.
- **Assess and Review** these approaches based on monitoring and analysis of the results achieved and ongoing learning activities. Finally, identify, prioritise, plan and implement improvements where required.

When using the model within a construction organisation, the Approach, Deployment, Assessment and Review elements of the RADAR logic should be addressed for each *Enabler* criterion and for each *Results* criterion.

EFQM's RADAR model mechanism is related to Deming's continuous improvement cycle (plan – do – think – act). See *Figure 4.2*. More importantly, the process is driven by self-assessment, which is not only a means for measuring continuous improvement, but also an excellent opportunity to integrate total quality management into normal operations (EFQM 2000b).

FIGURE 4.2 Deming's dynamic control loop cycle

Steps for implementing the EFQM Excellence Model

Ho (1999) summarised the critical steps for formulating and shaping a corporate strategy while undertaking the process of implementing the EFQM Excellence Model. The following steps relate to the relationship between a quality initiative and corporate strategy. Ho (1999) divided the corporate strategy into three key phases.

- The first phase is "the determination of a corporate mission statement which sets a common value for everyone in the organisation". Noting that a mission statement for an organisation is usually for a long-term period of at least 10 years.
- The second phase "is defining the strategic options and choosing the optimum one". This is the medium-term plan, which usually ranges from 3 to 5 years.
- The third phase is "the strategic implementation which is also known as operations management". This is the short-term plan which is usually three months to 1 year.

Having identified the above phases, Ho (1999) asks the question "where does this TQM initiative – EFQM Excellence model – fit into this Corporate Strategy?" In order to address this issue, it is best to consider quality as a routine organisational activity encompassed within the strategic planning and deployment process. The main advantage of this approach is summarised by Ho (1999) as "it adds totality to quality, as it is communicated throughout the [construction] organisation and spanned over its long term plan".

Moreover, Ho (1999) discusses the basis and rationale behind the success of TQM in Japanese companies. According to studies conducted on Japanese organisations in the manufacturing and services sectors, it has been found that "TQM was part of the daily language and activities in the organisations". The whole environment inside the organisations would reflect the embracing of quality. In other words, "TQM was integrated into the firms' management practices and operations [as does EFQM.E.M]".

Not only do policy and strategy need to be well planned and developed but they also need an inclusive approach if they are to be deployed at all levels within a construction organisation. Oakland et al. (2002) suggested using scorecards as a tool to be able to measure, review and update the policies during all stages, which will result in achieving a consistent approach to measuring progress towards the set organisational objectives.

People

Excellent organisations manage, develop and release the full potential of their people at an individual, team-based and organisational level. They promote fairness and equality and involve and empower their people. They care for, communicate, reward and recognise, in a way that motivates staff and builds commitment to using their skills and knowledge for the benefit of the organisation.

Oakland et al. (2002) suggest that the essence of the EFQM Excellence Model is its mechanism, dynamics and recognition of the importance of the employees' role, and that "processes are the means [through] which a company or organisation harnesses and releases the talents of its people to produce results performance".

Partnerships and resources

Excellent organisations plan and manage external partnerships, suppliers and internal resources in order to support policy and strategy and the effective operation of processes. During planning and whilst managing partnerships and resources, they balance the current and future needs of the organisation, the community and the environment.

Partnering

Partnering may be defined as a long-term commitment between two or more construction-related organisations for the purpose of achieving specific business objectives by maximising the efficiency and effectiveness of each participant's resources. This requires changing traditional relationships to a shared culture without regard to organisational boundaries. The relationship should be based on trust, dedication to common goals, and an understanding of each other's individual expectations and values. Expressed benefits of engaging in partnering include improved efficiency and cost effectiveness, increased opportunity for innovation and the continuous improvement of quality products and services.

In addition, partnering can be split into two main types:

- Project partnering is a partnering arrangement on a single project in which, at the end of the project, the partnering relationship is terminated.
- Strategic partnering is a partnership arranged on a long-term basis, in order to gain long-term synergistic advantages.

Experience has shown that strategic partnering provides more organisational benefits when compared with project partnering; this is because it allows time for continuous improvement and relationships to be developed over a sustained period of time.

Processes, products and services

Excellent organisations design, manage and improve processes, products and services in order to fully satisfy, and generate increasing value for, customers and other stakeholders.

Issues in process management

Orsini (2000) outlined the following steps for the achievement of process improvement:

- provide a clear definition of the processes under consideration;
- identify all the interactions of the processes under consideration with all other processes;
- produce a specification of the critical characteristics of the process under consideration;
- produce a means of measuring the critical characteristics of the process under consideration;
- collect sufficient and reliable data and determine whether the process under consideration is unstable or stable;
- establish the key performance indicators in order to be able to take advantage of any improvement opportunities that may develop; and
- be prepared to make changes (this can best be accomplished by using the RADAR approach).

"Failure to meet requirements in any part of a quality chain has a way of multiplying and failure in one part of the system creates problems elsewhere [thus the integrative nature of processes must be a prime consideration for companies]" (Orsini 2000). This concept is so significant that if senior management and employees can grasp it, it will ease the transformation of a cultural change and would thus lead to the accomplishment of a 'continual improvement culture', since this approach is a continuous process (Oakland et al. 2002).

Oakland et al. (2002) noted that in public sector organisations the process criteria consists of four elements

1   To provide direction and improvement.
2   To satisfy customer needs.

3   To manage and ensure the capability of the organisation.
4   To engage in the measurement function and improve organisational performance.

Processes integrated into one system

One of the fundamental problems associated with the deployment of any TQM/ EFQM.E.M activity is the failure to appreciate the integrative nature of the system. Orsini (2000) points out that if all employees worked as hard as they possibly could, it would not be sufficient to solve the problematic issues of a poorly designed organisational system.

In addition, Orsini (2000) describes that the condition of such an organisation will lead to poor communication and the ineffective implementation of corporate objectives. He summarises the key characteristics of such poor performance as:

*   inconsistencies in organisational procedures;
*   failure to properly manage people;
*   destructive internal competition;
*   sub-optimisation between groups, within a department, within a division, within a company, between a company and its suppliers, between a company and its customers;
*   a failure to think or plan ahead;
*   trying to engage in retrospective corrective actions 'after the fact', exhibited through audits, inspections and untimely feedback.

Table 4.2 provides a summary of key deployment issues along with the resulting advantages of deployment for construction organisations.

TABLE 4.2 EFQM Excellence Model deployment advantages for construction organisations

Key Deployment Issues	Resulting Benefits
• Process improvements	• A clear understanding of how to deliver value to clients and hence gain a sustainable competitive advantage via operations.
• Attaining an organisation's objectives	• Enabling the mission and vision statements to be accomplished by building on the strengths/ distinctive competence of the company.
• Benchmarking Key Performance Indicators (KPIs)	• Ability to gauge what the organisation is achieving in relation to its planned performance (Plan, Do, Check, Act). And engage in continuous organisational improvement.
• Development of clear, concise action plans resulting in a focused policy and strategy	• Clarity and unity of purpose so the organisation's people can excel and continuously improve.
• Integration of improvement initiatives into normal operational activities	• Interrelated activities systematically managed with a holistic approach to decision making resulting in a learning and improving culture.

Note: issues are not mutually exclusive.

Gradually re-framing the organisation

The EFQM.E.M provides a valuable framework for addressing the key operational activities of construction organisations. It is useful because it enables a link to be made between people, organisational objectives and improvement processes, all encompassed under the umbrella of continued improvement (EFQM 1999b).

Implementation effectiveness of an EFQM.E.M/TQM framework model can be improved through the gradual re-framing of an organisation. These improvements occur through determined phases; at the outset they should be well planned in order to address the expected consequences of cultural organisational change, which is never easy to achieve (Dooley 1998).

In addition, to be able to support a culture change as a result of the introduction of the EFQM Excellence Model, there has to be demonstrable senior management involvement. It is evident from previous documented experiences that holding regular team meetings could be a good means for securing senior management support. These meetings provide a communication link between senior management, employees and middle management. After all, it has already been established that one of the most important barriers to TQM (and hence EFQM.E.M) success is insufficient management support (Capon et al. 1995).

What is evident is that the EFQM.E.M does encompass all construction organisational activities, including their impact upon society, and provides a valid methodology for construction organisations to obtain a sustainable competitive advantage.

Culture

Culture is related to social anthropology and the study of "shared meanings and values held by groups in society that give significance to their actions" (McKenna 2000). In construction organisations, culture is often referred to in the context of 'corporate culture', which comprises "behaviour, actions and the values that people in an enterprise are expected to follow". While other researchers defined organisational culture as "a pattern of basic assumptions, invented, discovered, or developed by a given group as it learns to cope with its problems of external adoption and integral integration, that has worked well enough to be considered valid and, therefore, is to be taught to new members as the correct way to perceive, think, and feel in relation to those problems" (McKenna 2000).

Dooley (1998) addresses an important aspect of cultural change and the correct approaches to attaining such a sea change. There are two main approaches to be considered when deploying a change strategy, the first being a focus on an "individual's cognitive processes". The second approach relates to "their actions and interactions within the firm". These two approaches are referred to as *Thoughts* and *Action*.

The first approach is referred to as the 'Thoughts approach', where individual behaviour is attributed to knowledge, attitudes and beliefs. Change can occur through an organisation's training strategy, one that is supported by consistent, rational and coherent application. The second approach to change as noted is 'Actions', based on the assumption that an employee's knowledge, attitudes and behaviour are shaped by employee involvement (empowerment).

Changing the culture

Changing an organisation's culture is a vital component in the successful implementation of TQM/EFQM.E.M. Organisational change processes must be managed; first, the change of culture must be part of an explicit and comprehensive plan for improvement. Second, senior and middle management must be seen to be the driving force; in other words "managers must learn to lead change in an overt manner". Third, before starting to develop the plans for change, the views of managers and employees should be obtained and evaluated; the results of questionnaires, group meetings, etc. can provide the means for the information gathering process. Fourth, cultural change should be introduced as an ongoing process rather than a requirement of the introduction of TQM/EFQM.E.M. Fifth, to facilitate cultural change, construction organisations will benefit from the utilisation of management tools and techniques, used effectively and with purpose. TQM/EFQM.E.M application does not negate the use of management tools such as Statistical Process Control (Dale 1999).

In addition, Dale (1999) emphasises that employees' roles are a significant component of cultural change. Since employees are the real assets of an organisation, he proposed that the value of this asset (employees) would increase or decrease in accordance with the way employees are treated. Thus, Dale (1999) proposes that for a cultural change to be successful, the following should be considered: the fact that people have different backgrounds, ages, skills abilities, levels of enthusiasm, levels of competencies, levels of flexibility and ability to accept change.

Choosing and using the EFQM.E.M

Finn and Porter (1994) note that the foremost reason for using the EFQM Excellence Model was that it offered a "framework for exploring the link between organised activities and results and for driving continuous improvement". Further, Westlund (2001) identifies that the "EFQM Excellence Model has clearly become the most applied model in Europe for total quality management (TQM)".

Osseo-Asare Jr. and Longbottom (2002) propose that the nine criteria represent decisive success factors and encompass TQM principles. The application of the EFQM Excellence Model within a construction organisation can offer "a framework for exploring the link between organisational activities and a drive for continuous improvement". Curry (1999) summarises the uniqueness of EFQM.E.M by stating: "this model provides a tangible framework for assessing excellence in an organisation and for making step improvements in operations. It helps to bring greater cohesion to the different organisational activities".

In addition, Coleman and Douglas (2001) argue that the EFQM.E.M "defines and describes TQM in a way that can be more easily understood by senior management". This would allow them to accept ownership of any changes required and be able to drive their organisations towards excellence. It would also provide a tangible pathway to TQM, with clearly defined requirements. More importantly, Coleman and Douglas suggest that any lack of clear requirements for achieving TQM are not to be found in the EFQM Excellence Model.

Furthermore, there is another significant aspect that makes the choice of the EFQM Excellence Model much more appropriate than Baldrige's National Quality Award (MBNQA) and others. The EFQM Excellence Model was developed and based on lessons learned from previous experiences and other TQM models such as MBNQA. Organisations make use of self-assessment models as a tool to provide a path for "What they should do" (Ho 1999).

Osseo-Asare Jr. and Longbottom (2002) outline the main characteristics of the EFQM Excellence Model as follows:

• It provides a holistic way of managing a business enterprise, which will lead to its long-term success.
• The model is a diagnostic tool for self-assessment of the current health of an organisation. Self-assessment will provide the ability to balance an organisation's priorities, allocate resources and generate realistic business plans.

Pitt (1999) points out the reasons for choosing the EFQM Excellence Model over other models. First, the EFQM.E.M provides a broader approach to quality assurance and continuous improvement, in comparison to ISO 9001. Second, EFQM.E.M has been designed by European organisations; it can be said that the EFQM.E.M has a certain 'European Flavour', so it is easy to benchmark against other construction organisations across Europe. Third, the EFQM.E.M was developed and improved based upon other TQM models such as Malcolm Baldrige Model. Fourth, the model is capable of integrating other quality initiatives such as ISO 9001, Investors in People and the Charter Mark.

Furthermore, Ho (1999) suggests that an important reason behind choosing a particular TQM framework over another was the geographic location of the firm. There are three major Quality Awards, Japan's Deming Prize, USA's Malcolm Baldrige National Quality Award (MBNQA) and the European Quality Award (EQA). Therefore, organisations located in Europe, or which have a strong European presence would find the EFQM.E.M most appropriate. Choosing a model according to the location, allows for benchmarking with other organisations working within the same operational environment (Ho 1999).

Benchmarking

One of the key strengths of the EFQM.E.M is the self-assessment methodology, which forms a strong basis for benchmarking performance with other organisations. Benchmarking can be defined as "a positive, proactive process to change operations in a structured fashion [as part of a learning organisation culture] to achieve superior performance" (Ball et al. 2000).

Ball et al. (2000) attribute the success of benchmarking in the private sector to being utilised as a method for searching for new ideas and practices for adoption in a company and assisting in the attainment of competitive advantage. Benchmarking is used as a management tool for the improvement of organisational performance by engaging in organisational learning.

EFQM.E.M self-assessment processes

The European Foundation for Quality Management provides a useful definition of self-assessment for Quality Management (EFQM 2000b). Self-assessment is "a comprehensive, systematic and regular review of an organisation's activities and results against the EFQM Excellence Model. The Self-Assessment process allows an organisation [suitable for construction organisations] to discern clearly its strengths and areas in which improvements can be made. Following this process of evaluation, improvement plans are launched, which are monitored for progress".

One aspect of the EFQM.E.M is that it enables a reverse direction to self-assessment to be adopted. "This means adopting a diagnostic sequence, which starts from results (or symptoms)" (Conti 1997).

The first phase of self-assessment for construction firms consists of analysing results in order to establish performance gaps. Secondly scores may be allocated in order to pinpoint areas to be addressed, thus engaging in a continuous improvement process.

Dale (1999) advocated that self-assessment is an effective tool for achieving continuous improvement. He postulated that before embarking on self-assessment, both management and employees need to have some knowledge of self-assessment in order that they can understand the questions related to the self-assessment process.

Dale (1999) describes the mechanism of self-assessment as "one of the models underpinning an award to pinpoint improvement opportunities and to identify new ways in which to encourage the organisation down the road of business excellence".

"The concept of organisational learning involves teaching an organisation to use a scientific method (self-assessment), to create and utilise specific knowledge, and to change its performance measurement system" (Hendricks and Singhal 2001).

The EFQM.E.M, with its 'RADAR' application as an integral part of the model, does enable a positive change in organisational performance to be achieved, and thus continuous improvement is an attainable organisational objective.

Different approaches to self-assessment

There are different approaches to conducting a self-assessment process and these are now outlined; construction firms need to fully consider which approach is most suitable for them.

The award simulation approach

Although this is potentially the most time consuming and resource intensive of all the approaches described, it is very comprehensive. It will provide one of the most accurate scoring profiles allowing for legitimate comparisons with the scoring profiles of award applicants.

AWARD SIMULATION ADVANTAGES

- It produces a list of strengths and areas for improvement developed by a team of trained assessors which can be used to drive improvement actions.
- The process of writing the information down provides a powerful and concise way of reflecting the culture and performance of the construction company. As it is a written report, it can be referred to repeatedly.
- Once the first report has been completed, subsequent reports are relatively easy to complete with a high degree of accuracy and consistency.
- It provides an excellent opportunity for involvement and communication during the data gathering process. When completed it also provides an excellent communications document to be shared by the people within the company, its customers, suppliers and others with an interest in the organisation. Some construction organisations use the report as part of their marketing strategy.
- The site visit and presentation from the assessor team are important value adding steps, since they provide an opportunity to check deployment issues, and for the assessors to explain, in greater detail, the rationale behind their comments in the feedback report.
- The process provides a learning opportunity prior to application for the Quality Award for Excellence.
- It provides an easy way for units within a construction company to compare processes and results and identify examples of best practice that may be shared/disseminated throughout the organisation.

DISADVANTAGES

- There is a temptation for a management team to be less involved by taking the opportunity to delegate most of the work.
- It can be seen as an exercise in creative writing, covering up the real issues.
- For those construction organisations in the early days of their journey to Excellence, this approach may be too ambitious as a first attempt at self-assessment. This is because it is very resource intensive.

The ASSESS questionnaire approach

This is a comprehensive approach to self-assessment incorporating simple and easy to use questions and providing a focus on actionable data to aid improvement planning and benchmarking.

ASSESS QUESTIONNAIRE ADVANTAGES

- Questions are developed from the Model and so provide full coverage of the noted criteria for construction enterprises.
- A quick and easy entry point to self-assessment through Rapid Scoring, a more searching self-assessment through Team Scores, and an independently validated self-assessment.

- Scoring has been calibrated against the full Award process and the Model.
- The approach is software or paper based, or can be run in tandem.
- Each question is supported by hints and tips which provide more detailed insight into the questions.
- Is a very powerful learning approach to self-assessment for the individual and company.
- The software can capture narrative as well as the quantitative scores.
- The process can be undertaken individually or as a group.
- Individual assessments can be merged through the use of a powerful team facility in the software providing a focus for those areas that require consensus.
- Software provides export facilities to other reports and presentations.
- Has an excellent graphics facility.
- Output can be entered into a database for benchmarking purposes that will be compatible with the national benchmarking network.

DISADVANTAGES

- Needs a certain degree of software but can be paper based.
- Can be very searching of individuals and organisations as the questions are quite specific.

The pro-forma approach

Although the data gathering part of this process might be as long as the Award Simulation Approaches, the task of preparing the pro-forma, one page per part criterion is easier and less time consuming than drafting a full Award style report.

PRO-FORMA ADVANTAGES

- Provides a list of strengths and areas for development for driving improvement actions for construction enterprises.
- Allows people to document the evidence upon which strengths, areas for improvement and scores are based.
- Scoring profiles can be derived which, in terms of accuracy, lie closer to the Award simulation approach rather than, for instance, the matrix chart approach.

PRO-FORMA DISADVANTAGES

- The collection of pro-formas may not inform of the full story of the organisation. It represents a summary of the position.

The workshop approach

In terms of resources required, this approach does not take as long as the Award simulation process but on average is likely to take longer than either the matrix chart or questionnaire approaches.

WORKSHOP ADVANTAGES

- It is probably the best way to get a management team to understand and commit to the Model and its deployment.
- Discussion and agreement by the management team on the strengths and areas for improvement help to build a common and comprehensive view on the current state of the construction organisation. This leads to ownership by the management team of the output and facilitates their prioritisation and agreement of action plans.
- Provides a building opportunity for the management team.
- An agreed list of strengths and areas for improvement is produced which will drive improvement actions for all.

DISADVANTAGES

- A less robust and rigorous process than the Award simulation approach.
- Can be a high risk approach and needs excellent preparation and facilitation to ensure the management team are fully prepared and comfortable with the process. Ground rules for behaviour during the workshop should be agreed and understood beforehand.

The matrix chart approach

This approach is less resource intensive and quicker to use than the Award simulation approach provided an existing matrix chart is used. However, the resource and time requirements will increase considerably if a construction organisation chooses to create its own matrix chart. It is particularly suited for use by small teams.

MATRIX ADVANTAGES

- It is simple to use, as basic awareness training is sufficient to get things started.
- Can be used to involve everyone in self-assessment.
- Provides a practical way of understanding the criteria.
- Provides a means for teams to assess their progress quickly and easily and progress can be readily displayed. Gaps can also be clearly demonstrated, giving an indication of what to do next.
- Good for facilitating team discussions and team building.
- Involving the management team in developing its own matrix chart can be a powerful process and it forces them to discuss, reach consensus, articulate their collective vision and describe the steps towards achieving it in all nine criteria areas.

MATRIX DISADVANTAGES

- It does not provide an 'Award Standard' self-assessment; lists of strengths and areas for improvement are not produced.
- It does not allow for comparisons with Award applicants.

- There is not necessarily a direct link between the steps in the matrix chart and the criterion parts of the Model.

The peer involvement approach

This is similar to award simulation in terms of providing a comprehensive approach to self-assessment with the associated time and resource implications.

PEER INVOLVEMENT ADVANTAGES

- Less prescriptive than the Award simulation approach. The unit undergoing self-assessment does not have to produce a full report; the submission can be in any suitable form.
- Provides the opportunity for the involvement not only of people within the unit but also their colleagues from other parts of the construction organisation. This leads to a high degree of cross-functional learning for the assessors and the company.
- Provides a comprehensive list of strengths and areas for improvement for driving improvement actions, again for all.

PEER INVOLVEMENT DISADVANTAGES

- It can require the use of more resources than some of the other approaches.
- The degree to which units would volunteer for the exercise and be prepared to share information may limit the value of the exercise.

The simple questionnaire approach

This approach is one of the least resource-intensive and can be completed very quickly, provided an existing and proven questionnaire is used. It is an excellent approach for gathering information on the perceptions of people within the construction firm.

ADVANTAGES

- It is simple to use and some basic awareness training is sufficient to commence activities.
- It can readily involve many people within the company.
- Presentations of outcomes are less problematic.
- It is easy to compute and understand the numerical results.
- The questions asked can be customised to suit the construction company.
- It does provide a good introduction to self-assessment.
- Enables the organisation to receive feedback which can be segmented by function and level.
- It can be used to facilitate group discussions between teams on the opportunities for improvement within their units and as a total improvement process.

DISADVANTAGES

- Excessive use of questionnaires in any organisation may result in a low return. What response rate is a valid return?
- Not everyone in the company may understand the meaning of the questions.
- Wide circulation can raise expectations amongst the people within an organisation and the use of this approach will need careful positioning.
- Questionnaires tell you what people think, not why they think that way.
- A list of strengths and areas for improvement is not usually generated.
- Does not allow for comparison with scoring profiles of Award applicants.
- Accuracy and relevance depend upon the quality of questions asked.

(British Quality Foundation 1998)

In brief, Dale (1999) advocates that the process of self-assessment will constitute three main stages:

1 The gathering of data for each criterion related to the Model.
2 Conducting a valid assessment of the data gathered.
3 Developing appropriate plans and actions arising from the assessment and monitoring the progress and effectiveness of the plan of action.

Dale (1999) further outlines the key issues to be considered by organisations embarking on self-assessment:

- Ensure that senior management are committed to the self-assessment process and are prepared to use the results to develop improvement plans (not an easy task).
- Arrange for everyone involved in the process who requires some training to be trained.
- Communicate to all the rationale for engaging in the self-assessment process.
- Plan the means for collecting the data.
- Decide on the team and allocate roles and responsibilities for each criteria of the model.
- Develop a valid data collection methodology and identify data sources.
- Agree on an activity schedule and manage it as a project.
- Decide the best way for organising the data which has been collected.
- Present the data, reach agreement on strengths and areas of improvement and agree on the scores for the criteria.
- Prioritise the improvement and develop an appropriate action plan.
- Conduct regular reviews of progress against the plan.
- Repeat the self-assessment process as appropriate.

The uniqueness of the self-assessment processes lies in providing real evidence that can be utilised in the form of a trend analysis, thus enabling a construction organisation's momentum towards TQM/EFQM.E.M to be monitored and enhanced, encapsulated within the RADAR concept.

Using a combination of methods

Self-assessment methods based on workshops and pro-formas can involve relatively few people within an institution, although this is very much dependent on what the whole assessment process looks like. Questionnaires can provide extra data from a much wider base and thereby support either the pro-forma or workshop method.

A combination of pro-forma and workshops is useful. Pro-formas enable the gathering of a lot of detail and this – when carefully collated and presented – is an excellent basis for workshops, where the issues and supporting detail may be fully explored (EFQM 1999a, 1999b).

Benefits of using EFQM/self-assessment

Castka et al. (2003) note the benefits of using EFQM/self-assessment as:

1 Providing the opportunity to take a broader view on how the measured activity is impacting on the various business operations.
2 Measuring performance of processes, enablers and their relationship with organisational results.
3 Self-assessment conducted both internally and externally to the organisation.
4 Providing an opportunity to benchmark and compare like for like.
5 Measurement for providing improvement rather than for hard quality control.
6 Self-assessment is also an important communication and planning tool:
    6.1 The results of self-assessment provide a growing common language through which organisations, or parts of organisations, can compare their performances.
    6.2 The outputs of self-assessment are used for strategic management and action planning, or as a basis for an improvement project.
    6.3 New business values: leadership, people, process management, the use of information within the organisation and the way customer relationships are managed.

Underpinning the EFQM Excellence Model are the principles of knowing where a construction organisation is, where it wants to go and how it can get there. The model links self-assessment to informed planning and to implementation, through 'a framework of key processes'. Self-assessment can be seen as a catalyst for driving business improvement and hence achieving business goals.

Benefits derived from the implementation of the Excellence Model

Having recognised that corporate excellence is measured by an organisation's ability to both achieve and sustain outstanding results for its stakeholders, the enhanced version of the EFQM Excellence Model was developed. The fundamental advantages of the Excellence Model included:

- Increased cost effectiveness; results orientation; customer focus; partnership; knowledge management; performance and learning.

(European Foundation for Quality Management 1999a)

In a study on self-assessment, Hillman (1994) has elaborated further on the benefits of the EFQM Model, noting:

- It is not a standard, but it allows for interpretation of all aspects of the business and all forms of organisation.
- Its widening use facilitates comparison between organisations. This provides the potential to learn from others in specific areas by using a common language.
- The inclusion of tangible results ensures that the focus remains on real improvement, rather than a preoccupation with the improvement process, i.e. it focuses on achievement not just activity.
- Training is readily available in the use of scoring for the model.
- It provides a repeatable basis that can be used for comparison over several years.
- The comprehensive nature and results focus, when broken down into discrete elements, helps develop a total improvement process specific for each organisation – it is a model for attaining a successful business.

The following provide an underpinning rationale for construction companies to pursue a competitive strategy through the application of the EFQM Excellence Model. The Excellence Model is recognised as:

- providing a marketing focus;
- being a means of achieving a top quality performance in all areas of the firm;
- providing valid operating procedures for all staff;
- allowing for the review of organisational self-assessment performance through providing a competitive weapon via a quality approach; and
- incorporating the RADAR concept based upon the Deming improvement philosophy depicted in *Figure 4.2*.

EFQM.E.M deployment as a project

A generic deployment model is depicted in *Figure 4.3* (Watson and Seng 2001) and has been designed to be adopted or adapted in order to assist construction organisations when engaged in the implementation of the EFQM.E.M.

Watson (2002) describes the main differences between management and project management as being that management usually consists of a set of tasks that are repeated within a steady and reliable procedure. Project management, however, is related to the activities of a 'one off' specific project. A project can be more efficiently and effectively managed if the Deming dynamic control loop cycle is employed, as depicted in *Figure 4.2*.

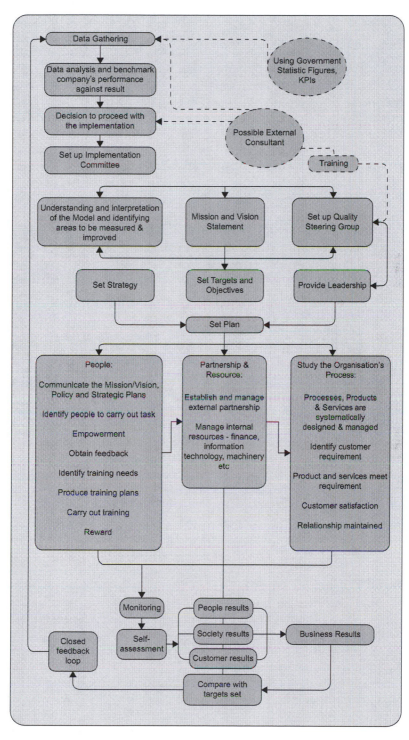

FIGURE 4.3   EFQM Excellence Model deployment

Harrison (1992) describes the traditional form of management as not being able to handle projects effectively. Watson (2002) emphasises that project management is a more challenging process. He described seven characteristics of project management. First, the role ends when the project ends. Second, start date and end date are difficult to predict due to the uniqueness of most projects. Third, the temporary nature of the project team. Fourth, many different skills are required when managing projects in teams. Fifth, costs can be very difficult to estimate. Sixth, often work has not been done before and is new to the team and unique to the project. Seventh, time, cost and quality constraints.

Harrison (1992) outlines specific characteristics of project management as:

1 projects are temporary [in nature and conducted] over a known duration;
2 they involve several departments and companies;
3 the complexity of integrating all activities, people and departments;
4 the organisation structure of a project is unique and complex;
5 different phases of the project [will] demand different [staff] groups for each phase, as a result relationships among staff [may become] unstable;

Furthermore, projects go through life cycle stages: conception, definition, design, execution, commissioning. The management of quality, time and costs is of prime consideration for the team and hence they play a key role in ensuring that each party involved in the project has well-defined objectives.

Kerzner (2001) suggests that to be an effective project manager, and for the deployment of EFQM.E.M (to be treated as a project), an individual must have management skills as well as technical skills. Kerzner (2001) proposes ten key skills of an effective project manager:

1 Team building skills: Managing cross functional teams and disciplines and being able to integrate them into one group.
2 Leadership skills: It is fundamental for a project manager to be able to manage and control the team. To do so a project manager must attain and demonstrate key personal characteristics such as "innovate thinking", "flexibility and change orientation" and "good communication skills".
3 Conflict resolution skills: In projects conflict is inevitable; the result of unresolved conflict among groups in projects may lead to delays to the project itself. However, conflict can be encouraged in certain conditions, where it may lead to innovations. The role of the project manager is to be able to resolve conflict, by generating an environment where the objectives of the project are clear, and identifying the causes of conflicts.
4 Technical expertise skills: It is important to be able to assess technical issues; however, a project manager should also use the skills, technical knowledge and competences of the team.

5 Planning skills: A project manager has to demonstrate the ability to plan and control all activities.
6 Organisation skills: Since the project manager's role is to integrate people from different departments, the project manager needs to understand how the organisation works and how to work with the organisation.
7 Entrepreneurship skills: The project manager must consider the broad vision of the organisation. There are many more issues to consider, for example, organisational growth.
8 Administration skills: A project manager should be familiar with the basic skills of "staffing", "budgeting" and "scheduling".
9 Management support: Usually a project manager is a linking pin between senior management and the rest of the project team members, who have to deliver the projects. Accordingly, project managers should be capable of building a good working relationship with all parties; however, they are also entitled to the support of senior management, should the occasion arise where it is required.
10 Resource allocation: A project manager needs to allocate the human and non-human resources in the most effective and efficient manner.

It should be appreciated that project managers have to manage up as well as down; this means they have to manage the client and senior staff of the organisation who may not be specifically involved in the project (the deployment of EFQM.E.M).

Effective teams in deploying the EFQM.E.M

Dale (1999) proposes that the 'health' of a team is a significant factor that needs to be regularly assessed. Dale also stresses that it is not an easy task to evaluate the effectiveness of teamworking. He suggests the following characteristics of an effective team:

• Everyone in the team is participating, making a contribution and is involved in appropriate actions and through this they are achieving their personal potential in line with project goals.
• Relationships are open and team members trust and respect each other.
• Members listen closely to the views of other members of the team and have an open mind and maintain a positive attitude towards the rest of the team and the project.
• Everyone can express their views and ideas with problems being addressed by the team, if appropriate.
• Members respect the operating procedures and principle of the team, and they own the team process.
• There is clarity and unity of focus on the project and all members know what is expected of them in achieving project goals.
• The TQM (EFQM.E.M) team leader has the ability to translate ideas into actions.

Watson (2000) notes the importance of teamwork as being the core element for attaining cultural change within an organisation. In addition, Watson (2002) emphasises that TQM (EFQM.E.M) is based on involving all employees in achieving the objectives of an organisation, personal objectives and maximising the performance of respective teams in the completion of project tasks. Personal objectives might include improved salaries and enhanced promotional prospects.

Love et al. (2000) discuss how employees can be formed into "self-directed work teams", which he defines as a number of employees forming an effective operational group. They suggest developing 'in-house employee education sessions' which consist of specific classroom teamwork exercises formed of multi-disciplinary groups (managers, office staff, etc.) in order to break down managerial and work sectional barriers. They also summarise the benefits of such workshop sessions as:

- Providing for an improvement in communication links between work section members.
- Assisting in breaking down work section barriers.
- Enabling an improved awareness of how teamwork can help in achieving and exceeding corporate and personal targets.
- Highlighting the significance and benefits of preparation and planning.
- Emphasising the need for effective time management.
- Underscoring the importance of effective resource management.
- Helping the development of an internal/external customer culture.
- Possibly assisting in the resolution of conflicts.

Ho (1999) addresses another important human factor within the TQM/EFQM.E.M implementation process; the establishment of 'Quality Control Circles'. Such circles are small groups of employees who actively contribute to improving the firm. They rely on using quality control methods to solve repeated occurring problems. Some of the methods utilised consist of cause-and-effect diagrams, histograms, scatter diagrams and quality control charts (Dale 1999).

Empowerment

Both individual and team empowerment are essential components in the successful application of the EFQM.E.M. Individual empowerment is vital for developing the self-esteem of the employee and to encourage individual motivation. While team motivation helps groups to deal more easily with difficult business environments, it also leads to the effective integration of group/team members (Dainty et al. 2002).

Dainty et al. (2002) suggest that empowerment comprises three core elements:

- Clarity of vision and mission.
- Consistency of organisational goals and the alignment of systems directed toward these goals.

- An ongoing evaluation of the professional needs of employees, and congruence between corporate, management and employee goals.

Several benefits can be attained as a result of attaining effective empowerment in individuals and teams:

- Improved productivity and quality.
- Reduced operating costs.
- Greater flexibility.
- Increase in job satisfaction and hence the improved motivational aspects of all staff.

Teamwork through self-assessment

It has been established that the involvement of all employees is one of the fundamental components of TQM and hence a key component of the EFQM.E.M. The most appropriate method of involving all employees in a systematic review of their processes is through self-assessment application. The process of self-assessment results in the identification of an organisation's strengths and potential improvement opportunities (Finn and Porter 1994).

Training

Dooley (1998) emphasises the importance of training as a way of improving the implementation of TQM/EFQM.E.M, noting that organisations must introduce and provide major improvements in TQM/EFQM.E.M training and educational programs. Training needs to be well organised and clearly focused on the needs of the company and project demands. Dooley (1998) proposes that training should be practical and sometimes even experimental and most importantly that the learned skills and competences must be implemented immediately in order for transfer of learning to occur. Further training programmes should develop specific objectives for learning, so assessment could be conducted by measuring how successful the training programme is against the determined objectives.

Another method that can assist in the effective application of TQM/EFQM.E.M is based on the establishment of a "broad networking with other firms". This can be accomplished in various ways such as conferences, monthly meetings and professional body activities, such as Continuing Professional Development (CPD) events. An obvious way of gaining new skills quickly for a construction company is to seek out and employ external consultants.

Training programmes must be consistent in content if offered throughout an organisation and should take into consideration any possible cultural differences that may exist between departmental boundaries (Reavill 1999).

Conclusion

The EFQM Excellence Model provides a valuable framework for addressing the key operational activities for construction related organisations. It is useful because it enables a link to be made between people, organisational objectives and improvement processes, all encompassed under the umbrella of a continuous improvement philosophy.

Many construction organisations suffer from poor performances because of a combination of traditional organisational structures and management practices, while operating within a dynamic competitive environment. Love et al. (2000) advocate that TQM/EFQM.E.M could be the solution for such construction organisations by "implementing the philosophical elements of total quality management".

Watson (2000) states that "the EFQM Model provides a truly service focused quality system which has an inbuilt mechanism for the attainment of continued organisational improvement". Van der Wiele et al. (1997) identify that "the criteria of the model helped managers to understand what TQM means in relation to managing a company".

The application of the model is simple, holistic, dynamic and flexible. Further the model enhances senior management's understanding of TQM (Watson and Seng 2001).

The EFQM Excellence Model is used as a tool for assisting in defining TQM in a way which construction senior management and employees can comprehend and apply. However, to use any self-assessment method effectively, various elements and practices have to be in place and management needs to have had some experience of TQM to understand the questions underpinning the model on which self-assessment is being based (Van der Wiele et al. 1997).

To be able to utilise the EFQM Excellence Model and conduct self-assessment optimally, there must be a trend for "serious investment in resources" (Sommerville and Robertson 2000). The most essential resource is people, since they are the real asset of any construction organisation. Wright (1997) pointed out: "Indeed, it [EFQM.E.M] recognises that satisfied customers and staff are a far more powerful indicator of sustainable future success than financial measures alone". Furthermore, Dale (1999) suggests that the performance of the organisation increases or decreases according to the way employees are treated and deployed. A "total quality approach is built on the commitment and motivation of people, achieved through relevant training, good communication and genuine consultation, all features of any effective [construction] organisation".

This chapter has outlined the European Foundation for Quality Management Excellence Model, and the advantages of its deployment have been discussed. Construction organisations have much to gain by applying the EFQM.E.M.

Questions for the reader

Here follows some questions related to the information presented within this chapter. Try to attempt each question without reference to the chapter in order to assess how much you have learned. The answers are provided at the end of the book.

Question 1

The European Foundation for Quality Management (EFQM) has stated that the functions of their Excellence Model may be split into four components. Identify these four component parts. The answer is provided at the end of this text book.

Question 2

The advantages of utilising EFQM.E.M's self-assessment methodology have been noted by Castka et al. (2003). Identify the advantages of EFQM.E.M's self-assessment methodology.

Question 3

The EFQM.E.M is based and supported by specific concepts which are referred to as "The Fundamental Concepts of Excellence". Identify the noted Fundamental Concepts of Excellence.

Question 4 – case study

Deploying EFQM.E.M at XYZ construction company

A new managing director has just been appointed to XYZ; the appointment has been made on the understanding that she will oversee the deployment of the EFQM.E.M within the company. However, the managing director has only a limited knowledge of the model. Yet she has to convince all company personnel of the deployment rationale. Therefore she has decided to engage external consultants to assist her. You have been appointed as external consultant and asked to prepare a presentation for the board of directors (based upon this chapter). Your presentation should consist of key bullet points. The bullet points should relate to the advantages of deploying the EFQM.E.M; however, you should also note any possible problematic issues of implementation.

Question 5

Identify and list the nine key enabler and results criteria of the EFQM.E.M.

Further reading

British Quality Foundation (1998). *Guide to the Business Excellence Mode: Defining World Class*. London: British Quality Foundation.
Crosby, P.B. (1979). *Quality If Free: The Art of Making Quality Certain*. New York: McGraw-Hill.
Hermel, P., and Ramis-Pujol, J. (2003). An evolution of excellence: Some main trends. *TQM Magazine*, 15 (4), pp. 230–243.

Powell, T.C. (1995). Total Quality Management as competitive advantage: A review and empirical study. *Strategic Management Journal*, 16, pp. 15–37.

Watson, P. (2002). Developing an Efficient and Effective Control System. *Journal of the Association of Building Engineers* (February), 77 (3), pp. 28–29.

Wilkinson, A., and Witcher, B. (1991). Fitness for use? Barriers to full TQM in the UK. *Management Decisions*, 29 (8), pp. 46–51.

References

Ball, A., Bowerman, M., and Hawksworth, S. (2000). Benchmarking in local government under a central government agenda. *Benchmarking: An International Journal*, 7 (1), pp. 20–34.

British Quality Foundation (1998). *Guide to the Business Excellence Mode: Defining World Class*. London: British Quality Foundation.

Capon, N., Kay, M., and Wood, M. (1995). Measuring the success of a TQM programme. *International Journal of Quality & Reliability Management*, 12 (8), pp. 8–22.

Castka, P., Bamber, C., and Sharp, J. (2003). Measuring teamwork culture: The use of a modified EFQM model. *Journal of Management Development*, 22 (2), pp. 149–170.

Cherkasky, S.M. (1992). Total quality for a sustainable competitive advantage. *Quality*, August, 31 (8), pp. 4–7.

Coleman, S., and Douglas, A. (2001). Where next for ISO 9000 companies? *Total Quality Management Journal*, 15 (2), pp. 88–92.

Conti, T. (1997). Optimizing self-assessment. *Total Quality Management*, 9 (2 & 3), pp. S5–S15.

Curry, A. (1999). Innovation in public service management. *Managing Service Quality*, 9 (3), pp. 180–190.

Dainty, A., Bryman, A., and Price, A. (2002). Empowerment within the UK construction sector. *Leadership and Organization Development Journal*, 23 (6), pp. 333–342.

Dale, B. (1999). *Managing Quality*. 3rd ed. Oxford: Blackwell Publishing.

Dooley, K. (1998). Perceptions of success and failure in TQM initiatives. *Journal of Quality Management*, 3 (2), pp. 157–174.

European Foundation for Quality Management (1999a). *Radar and the EFQM Excellence Model*. EFQM Press Releases & Announcements. [Online]. Available at: www.efqm.org/seflas.htm. [Accessed 12 June 2000].

European Foundation for Quality Management (1999b). *Eight Essentials of Excellence: The Fundamental Concepts and Their Benefits*. Belgium: Brussels Representative Office.

European Foundation for Quality Management (2000a). *History of the EFQM*. Available at: www.efqm.org/history/htm. [Accessed 12 June 2000].

European Foundation for Quality Management (2000b). *EFQM and Self-Assessment*. Available at: www.efqm.org/seflas.htm. [Accessed 12 June 2000].

Finn, M., and Porter, L. (1994). TQM self-assessment in the UK. *The TQM Magazine*, 6 (4), pp. 56–61.

Harrison, F.L. (1992). *Advanced Project Management, a Structured Approach*. 3rd ed. Aldershot: Gower Publishing Company Ltd.

Harrison, M. (1993). *Operations Management*. London: Pitman Publishing.

Hendricks, K.B., and Singhal, V.R. (2001). Firm characteristics, Total Quality Management, and financial performance. *Journal of Operations Management*, 19, pp. 269–285.

Hickman, C.R., and Silva, M. (1989). *Creating Excellence*. London: Unwin Hyman Ltd.

Hillman, G.P. (1994). Making self-assessment success. *Total Quality Management*, 6 (3), pp. 29–31.

Ho, S. (1999). From TQM to business excellence. *Production Planning and Control*, 10 (1), pp. 87–96.

Kanter, R.M. (1989). *When Giants Learn to Dance, Reading.* New York: Touchstone.

Kerzner, H. (2001). *Project Management: A Systems Approach to Planning, Scheduling, and Controlling.* 7th ed. Hoboken, NJ: John Wiley & Sons, Inc.

Love, P., Li, H., Irani, Z., and Faniran, O. (2000). Total Quality Management and the learning organisation: A dialogue for change in construction. *Construction Management and Economics*, 18, pp. 321–331.

McKenna, E. (2000). *Business Psychology and Organisational Behaviour.* 3rd ed. London: Psychology Press Ltd.

Mundy, K. (1992). Making the right choice. *Logistics Today*, September–October.

Nunney, D. (1992). *Integrated Manufacturing.* The Department of Trade and Industry UK.

Oakland, J., Tanner, S., and Gadd, K. (2002). Best practice in business excellence. *Total Quality Management*, 13 (8), pp. 1125–1139.

Oakland, S.J. (1993). *Total Quality Management.* London: Butterworth Heineman.

Orsini, J. (2000). Troubleshooting your activities for excellence. *Total Quality Management*, 11 (2), pp. 207–210.

Osseo-Asare, A. Jr., and Longbottom, D. (2002). The need for education and training in the use of the EFQM model for quality management in UK higher education institutions. *Quality Assurance in Education*, 10 (1), pp. 26–36.

Pearson, A. (2002). Don't go KPI nuts. *Building*, 267 (5), pp. 44–45.

Pitt, D. (1999). *Improving Performance through Self-Assessment.* NHS Trust, Wakefield, UK: Business Development Manager, Wakefield and Pontefract Community Health.

Preston, J. (1993). *International Business.* London: Pitman.

Reavill, L. (1999). What is the future direction of TQM development? *The TQM Magazine*, 11 (5), pp. 291–298.

Reed, D. (1998). *Self Assessment Techniques for Business Excellence: Identifying Business Opportunities.* London: British Quality Foundation.

Ross, D. F. (1991). Aligning the organisation for world-class manufacturing. *Production and Inventory Management Journal*, 32 (2), p. 22.

Sommerville, J., and Robertson, H.W. (2000) A scorecard approach to benchmarking for total quality construction. *International Journal of Quality and Reliability Management*, 17 (4/5), pp. 453–466.

Spencer, B.A. (1994). Models of organisational and Total Quality Management: A comparison and critical evaluation. *Academy of Management Review*, 19 (3), pp. 446–471.

Van der Wiele, A., Dale, B.D., and Williams, A.R.T. (1997). ISO 9000 series registration to Total Quality Management: The transformation journey. *International Journal of Quality Science*, June, 2 (4), pp. 236–252.

Watson, M. (2002). *Managing Smaller Projects.* 2nd ed. Swindon: Project Manager Today Publications.

Watson, P. (2000). Applying the European Foundation for Quality Management (EFQM) Model. *Journal of the Association of Building Engineers*, 75 (4), pp. 18–20.

Watson, P., and Chileshe, N. (2001). *The Relationship between Organisational Performance and Total Quality Management within Construction SME's.* Proceedings of CIB World Building Congress, April 2–6, Wellington, New Zealand (3), pp. 233–244.

Watson, P., and Seng, L.T. (2001). Implementing the European foundation for quality management excellence model in monstruction. *Construction Information Quarterly*, 3 (2), p. 130.

Westlund, A. (2001). Measuring environmental impact on society in the EFQM system. *Total Quality Management*, 12 (1), pp. 125–135.

Wright, A. (1997). Public service quality: Lessons not learned. *Total Quality Management*, 8 (5), pp. 313–322.

Yip, G.S. (1992). *Total Global Strategy*. Englewood Cliffs, NJ: Prentice Hall.

5

DEVELOPING ORGANISATIONAL LEARNING

Project and corporate learning, linked to continuous improvement, provides the focus for this chapter. The chapter proposes that in order for construction organisations to fully engage in a continuous improvement process and strive for competitive advantage, they must develop the culture of a learning organisation. It is suggested that the concept of organisational learning should be linked to the key functions of management; functions that serve to control organisation resources, procedures and systems. A self-assessment model that considers the management functions of construction organisations is outlined. This self-assessment model – the Management Functional Assessment Model (Watson 2005a) – serves to enable continuous improvement and excellence when linked with RADAR.

Learning outcomes

By the end of this chapter the reader will be able to demonstrate an understanding of:

- The rationale for developing an organisational learning culture.
- The basic requirements for developing a learning organisation.
- The advocated advantages and cultural values that underpin organisational learning.
- The Management Functional Assessment Model and how it can be applied to enable organisational improvement and continuous improvement.

Introduction

This chapter explores the utilisation of a self-assessment methodology based upon the key management components of construction firms. The building blocks of the management process are defined and are linked to organisational learning.

In order for construction organisations to fully engage in a continuous improvement process and strive for competitive advantage, they must develop the culture of a learning organisation. The concept of organisational learning should be linked to the key functions of management; functions that serve to control organisation resources, procedures and systems.

A Management Functional Assessment Model (MFAM) provides the means for construction organisations to gauge the effectiveness and efficiency of their management activities and provides a means for the attainment of organisational continuous improvement.

The MFAM model embraces the seven functions of management and overcomes the critical issue noted by Greising (1994):

> that it is too easy for managers to become overly enamoured with the procedures and mechanisms for TQM while forgetting that the point of the activity is to improve firm performance and that quality can go up but profits can go down.

The model encapsulates all management functions, as it is designed to empower improvements in all aspects of project and corporate performance and attain competitive advantage. This model is considered suitable for all sized businesses in both service and manufacturing sectors.

The model is underpinned by the premise that strategic advantage does not come from the simple possession of assets or of particular product/market position, but from a collection of attributes which are built up over time. These attributes provide a basis for achieving and maintaining a sustainable competitive edge in an uncertain and rapidly changing dynamic operational environment.

Continuous improvement can be considered an example of what many strategy theorists call "dynamic capability" (Teece and Pisano 1994). Three elements are normally considered to constitute dynamic capability: paths; position; and processes (Tidd et al. 1997). The first two concern the amalgam of competencies that the construction organisation has accumulated and the particular position that it is able to adopt in its product/market environment. However, the third is of particular interest as it concerns the specific behavioural routines which characterise "the way we do things in this organisation" and which describe how the construction company approaches issues of innovation, learning and improvement.

Continuous improvement represents an important element of any dynamic capability, since it offers mechanisms whereby a high proportion of the organisation can become involved in its innovation and learning processes (Bessant and Caffyn 1997; Bessant 1998; Robinson 1991; Schroeder and Robinson 1993). It corresponds to what is widely known as "kaizen" and forms an important component of the "lean thinking" approach adopted by many construction companies (Imai 1987). Its strategic advantage is essentially as a cluster of behavioural routines but this also explains why it offers considerable competitive potential, since these behaviour patterns take time to learn and institutionalise, and are hard to copy or transfer. The potential for continuous improvement to become an enabling mechanism in organisational learning has been advocated by Nonaka (1991) and Leonard-Barton (1992).

Human resource development in construction has previously concentrated on developing an individual's skills and knowledge related to task, rather than corporate learning concepts.

The need for organisations to become learning companies has been asserted by a number of authors as a response to the increasing organisational challenges posed by rapid environmental change, discontinuity, economic uncertainty, complexity and globalisation. Indeed, one reason for the growth in popularity of the term is that it seems to capture many of the qualities deemed necessary for contemporary organisations such as teamwork, empowerment, participation, flexibility and responsiveness. Stata (1989) argues that "the rate at which individuals and organisations learn may become the only sustainable competitive advantage". The above noted components of becoming a learning organisation are impacted upon by the management functions of construction companies and these are now explored in more detail.

The seven functions of management

The task of managers may be summarised as having responsibility for forging into a holistic whole the three constituents of people, ideas and things. The attainment of this demanding task is assisted by addressing functions of management identified by Fayol (1949) and later extended. Functions of management include:

- Controlling
- Planning
- Forecasting
- Organising
- Motivating
- Co-ordinating
- Communication.

Overlapping and running through the above six functions is the seventh key function of 'communication'. This is the life-blood of any construction organisation and without which a construction manager cannot function efficiently or effectively.

An outline of the seven functions of management is presented for the reader; these outlines are provided in order to establish their importance, and their impact upon organisation quality initiatives.

Controlling

Control is concerned with the effective and efficient utilisation of resources in the attainment of previously determined objectives, contained within a specific identified plan. This plan may take many forms, e.g. Bar Chart, Network Analysis or a Financial Plan such as a Project Budget Plan – the plan being the method that requires deployment in order to achieve the pre-determined set objectives. However, it should also be based upon the most efficient and effective way of completing the set task(s).

Control is exercised by the feedback of information upon actual performance when compared with the pre-determined plan; therefore planning and control are very closely linked. Control is the activity which measures deviations from planned activities/objectives and further initiates effective and efficient corrective actions via a feed forward mechanism.

In order to have both efficient and effective control, the Deming dynamic control loop cycle should be employed as depicted in *Figure 4.2* of Chapter 4 (p. 104) within this text book. It is important that any information contained within the loop must:

- separate information according to areas of responsibility and accountability;
- present results in a consistent, readily understood and useful manner;
- represent appropriate and valid time periods for instigating effective actions;
- be available in a timely manner enabling effective decisions to be taken;
- divert the minimum energies from corporate primary functions, considering the 'Law of Diminishing Returns' and associated 'Opportunity Costs'; and
- demonstrate clearly the deviations from the pre-determined plan and the impact of noted deviations (if at all possible).

The above can be encapsulated under the two headings of 'cycle time', how long it takes for the information to circulate, and 'quality of information', the level of detail encapsulated in circulation (loop) processes.

For effective and efficient control, one must have short (appropriate) cycle times and the level of detail necessary (quality of information) to make valid decisions and deploy appropriate actions.

Planning

Planning is that aspect of management concerned with the particular rather than the general and is dependent upon the attainment of both reliable and accurate information. All construction managers plan, set objectives and try to anticipate the future in order to achieve set tasks.

Construction managers determine the broad lines of operations and the strategy or general programme, choose the appropriate methods, and sometimes the materials and machines required for the most effective and efficient action. So planning relates to how, when and where work is to be carried out.

The process of planning usually refers exclusively to those operations concerned with, and the department responsible for, determining the manner in which a job/project is to be executed along with the necessary resources.

The word 'planning' in the sense of forethought can also include such varied activities as market research, training schemes and the recording of plant locations and availability.

To be really effective, planning must be simple, flexible, balanced and based upon accurate standards of performance determined by systematic analysis of observed and recorded facts.

Planning is perhaps one of the most important tools of management, requiring intense application, precise attention to detail, imagination and a sound knowledge of technical theory, but is always a means, and not an end in itself.

In its application, construction managers should give full regard to the human needs of the organisation (covered further under the heading of Motivation).

Forecasting

Forecasting or looking ahead is generally the prerogative of senior managers, although it can enter into the decision-making process at any organisational level.

The only reliable method of arriving at important policy decisions is by the adoption of a systematic approach based upon a precise diagnosis of the situation, the collection and tabulation of all the facts, a dispassionate consideration and prognosis and the formulation of a logical conclusion.

Forecasting objectives can be as varied as economic forecasting, i.e. how much capital is required and which is the best source, estimating margins for tenders, the alternatives of buying and hiring plant or the selection of appropriate personnel.

Information may be in the form of trend analysis indicated by statistical control figures, market research results or even the feedback on recruitment interview tests.

The final outcome of the forecasting process may involve things like setting tender prices, determining estimated labour requirements or even staff promotions.

Competent direction is an essential factor in efficient and effective management and requires the qualities of broad vision, clear and incisive thinking, courage, self-confidence and a good judgement of personnel and situations.

Consideration of all factors involved is of particular importance when forecasting, as is the investigation of all possible alternatives, for the plan selected will most certainly highlight any ill-considered or uninvestigated areas and items. The planning and forecasting functions are very closely related. Fryer (1997) argues that long-range 'planning' is really 'forecasting' and the authors of this text book agree.

Organising

Organising is the other aspect of management activity which is complementary to planning and concerned with the more general selection of the people and the operational methods necessary for the discharge of managerial responsibilities. Construction managers are organising when they commence putting plans into action.

The process of organising or preparing comprises:

- the definition and distribution of the responsibilities and duties of the various management and supervisory personnel forming the establishment of the enterprise;
- the recording of the types of formal relationships existing between individual appointments, the pattern of accountability and theoretical paths of contact; and
- the formulation and deployment of standard procedures, preferred methods of working and operating instructions for standard techniques.

Certain guiding principles can be used by construction managers in order to determine the organisational structure; these include:

- Schedules of responsibilities, the organisational chart and standard procedures, should preferably be written down and distributed so far as applicable for general reference purposes, to allow revision and to preserve continuity despite transient personnel.

• When increasing size dictates the sub-division of responsibilities, this should be determined by functional or operational specialisation.

• Where possible, a single head person should be responsible to the policy forming body for the implementation of all the operations of the business.

• Decentralisation of decisions should be provided by the adequate delegation of responsibility, and any limitations should be specifically noted, management by 'Exception' could be employed here.

• Clear lines of accountability should link the chief executive with all points of the organisation, and the integration of specialists should not interrupt the organisational lines of command and communication.

• The structure must be flexible enough to facilitate amendments when circumstances change, but since endurance is the ultimate test of an enterprise's success, a formal outline and value system are necessary to assure the continuous and effective functioning of the construction company.

• A typical construction organisation does not exist, since consideration must be given to the individual characteristics and operational environments of each undertaking.

Successful construction managers divide up the total operation into individual jobs/tasks in order to be able to match them to correct personnel. However, they still have to co-ordinate them, so that one work group is not held up by another and to ensure that materials are there when required. "The function of organising is very specific to the [construction] manager's role" (Fryer 1997).

Motivating

Many authors have argued that an organisation's most important asset, particularly in a labour intensive industry like construction, is its people (Fryer 1997).

Motivation is a very complex function of management and there exists a wealth of published information on this topic by well-known management experts.

It is important to provide an acceptable definition of motivation and Cole (1995) provides one: "Motivation is the term used to describe processes, both instinctive and rational, by which people seek to satisfy the basic drives, perceived needs and personal goals, which trigger human behaviour".

It may be postulated that there are fundamentally two main types of motivation to work. One is the job as an end in itself (intrinsic satisfaction); the other is the end towards which the job provides the means (extrinsic satisfaction).

Intrinsic satisfaction

This is derived by fulfilling your own motivational needs from the job and is therefore achieved from work itself. A considerable weight of behavioural scientific research has been devoted to the pursuit of this concept.

Extrinsic satisfaction

This is deriving satisfaction of needs using work as a means to an end. It is some-times termed the 'instrumental approach'. Work provides us with money, and money enables us to 'buy' satisfaction to a certain extent, thus pay acts as the main motivating factor.

To try and draw some conclusions from the two schools of thought it could be stated that people work for different personal reasons but, basically, they fall into two categories: extrinsic and intrinsic. How can a construction manager motivate a workforce which is most likely to be a mixture of the two catego-ries? For the extrinsic workers it would appear that financial stimulus is the only means of motivation. This tactic has been tried in the form of incentive payments and bonus systems and indeed these usually do work for this type of employee. The opportunity to earn more is taken up and production increases (though this tends to be only a short-term phenomenon). What can managers do about their intrinsically motivated employees, who may never be motivated by money alone?

Intrinsic workers require motivation from the task and in order to achieve a desirable state of high morale it is necessary to:

- Arouse interest by keeping everyone informed of proposed developments and the progress of activities/set tasks.
- Foster enthusiasm by assisting in the attainment of legitimate personal and social satisfaction.
- Develop harmony and a sense of participation by engaging in joint consulta-tion processes.
- Enlist co-operation by providing reasonable continuity of employment and security for their future.
- Secure loyalty by showing fairness and being consistent in the allocation of duties, distribution of rewards, administration of discipline, etc.
- Promote keenness by fostering a sense of competition (where appropriate) and group or personal achievement.
- Encourage self-discipline by developing a sense of responsibility and the enjoy-ment of trust.
- Inspire confidence and respect by fair judgement and impartial dealings with subordinates.
- Ensure acceptance of the necessary rules and regulations by inspiring a sense of duty and a responsibility for the affairs of the construction organisation and individuals.
- Assist ambition by the encouragement and the affording of opportunities for individual development.
- Prevent frustration by providing a sympathetic and effective outlet for griev-ances and misunderstandings.
- Provide timely feedback on performance.

   Motivation is a vitally important concern to both employees and managers within an organisation. Its importance arises from the simple but powerful truth that poorly motivated people are likely to perform poorly at work and gain little satisfaction from their job.

(Naoum 2001)

Co-ordinating

This is the linking together of the various members to constitute a practical ensemble and the balancing of resources and activities to ensure as far as is practicable the complete harmony of processes and performance.

The main aims of the co-ordination function are first to ensure the prevention of separation of activities into watertight compartments as a sequence of specialisation, and secondly the preservation of a recognisable unity throughout the enterprise. In other words, to ensure a truly holistic approach to all construction organisational activities, bounded by a set of common goals.

A tendency for organisational activities to separate into individual functions increases with the size of the company, therefore keeping a functional team together may become a vital task for a construction manager.

Deliberate co-ordination of management may require specified activities such as regular meetings to integrate ideas and actions, the establishment of an additional effective communication system, and possibly for the attainment of greater clarity a pictorial presentation of responsibilities to assist co-operation between individuals and teams.

The performance of successful co-ordination requires:

- early introduction of the function;
- direct personal contact with all parties concerned;
- a reciprocal activity by the personnel being co-ordinated; and
- continuous operation of the function and the monitoring of its effectiveness.

Co-ordination is achieved in the main by the efforts and skill of the individual construction manager with due regard to the overriding human factors involved. It is the assessment and utilisation of core skills in the best interests of the organisation and its employees, with particular emphasis being placed on the harmony of major resource elements.

A construction manager has to co-ordinate a very diverse range of resources; these are the 5 M's: materials, manpower, management, money and machines. They must also integrate the work of subcontracts, all within a specific time frame and directed towards the achievement of set project/corporate tasks.

Communication

Communications are the means employed by executives to pass on their plans and instructions for action and by managers to make known their objectives/requirements and to inspire the necessary efforts, and by supervisors to co-ordinate activities and control operations.

Communication is a means of achieving contact between departments and individuals, and a channel for the distribution of knowledge; these are obvious fundamental activities of construction management. However, equally important aspects of communicating are not always fully appreciated by construction managers and these incorporate the following:

Good communications promote a better understanding by describing what is being done and why, and permits free expression of suggestions by all levels of personnel. This encourages a sense of participation and prevents friction and misunderstandings, thus contributing to the achievement and maintenance of healthy staff morale.

Although the process of communicating is the indispensable 'tool' of management or supervision, nevertheless the ability to convey messages clearly, vividly and convincingly, by either speech or writing, is the key to the exercise of power. The spoken word is more infectious and particularly useful for short-term persuasion, while the written word is more permanent and hence usually more suitable in the long term.

To be effective both arts necessitate the possession of accurate facts, the use of simple, precise language and the fluency of expression developed via constant practice by construction managers.

It should be remembered that "[m]any [construction] organisational problems are caused by communication failure. Breakdowns occur because of faulty transmission and reception of messages and because people put their own interpretation on what they see and hear" (Fryer 1997).

Communication with employees can be defined as the passing on and receiving of signals from one human being to another. (Of course computers can now serve this purpose.) The purposes of communication are threefold:

- to increase knowledge and/or understanding; the construction manager may not want the workforce to change their behaviour;
- to influence or change attitudes, although direct verbal communication may be designed with a view to changing attitudes, it is unlikely to do so; and
- to instigate or influence action or behaviour; ultimately all communication, particularly in the workplace is perhaps geared to this end.

Although construction managers may seek to increase knowledge or change attitudes, they will only have proof of having done so if behaviour of the recipient/teams changes as a result.

Overview of functions

Construction managers' jobs are demanding, complex and varied; however, there are a set of common features in the role of these managers, but individual jobs do differ considerably. Thus the amount of time and the way construction managers perform the previously noted functions of management depend very much on their abilities and personal motivation, and the motivation of their teams and team members.

In the 1990s there was a growing interest in empowerment, a process which shifts some of the power from the managers to employees, individually and as self-managed teams. Employees, being closer to the workforce and having a superior knowledge (in some cases) of the work and its environment, are often in a better position to make decisions; empowerment gives them the opportunity to use and develop their talents more fully.

(Fryer 1997)

As the above quote noted, there is a fine line between management delegation and abdication. Empowerment should be a shared joint experience/venture between construction managers and their employees. It is not and should not be an excuse to abdicate a construction manager's responsibility. Thus the activities/functions described in this section are still valid if a construction organisation is to be competitive and continually improve.

Construction managers can also act as catalysts for change within the context of organisational culture.

The culture of an organisation is its customary and traditional way of thinking and of doing things, which is shared to a greater or lesser degree by all its members, and which new members must learn, and at least partially accept, in order to be accepted. . . . [It] covers a wide range of behaviour: the methods of production; job skills and technical knowledge; attitudes towards discipline and punishment; the customs and habits of managerial behaviour; the objectives of the concern; its way of doing business; the methods of payment; the values placed and different types of work . . . and the less conscious conventions and taboos.

Jacques 1952 (cited by McCabe 2001)

Senior managers can clearly affect organisational culture, and organisational culture can assist or obstruct organisational change management processes; when considering organisational change processes senior construction managers must demonstrate that they are:

capable of driving cultural change through various organisational levels; [they must also have]

• Leadership skills
• Motivation skills
• Skill in dealing with resistance
• Skill in recognising different approaches, values and norms [impacting on culture]. . . .

The sort of person appointed to manage the implementation of culture change will need to be someone who, as well as being absolutely competent in technical issues surrounding the subject, possesses abundant confidence and human relation skill.

(McCabe 2001)

The critical importance of the seven functions of management has been outlined. It is clear that in order to manage quality and improvement processes a construction company must understand the seven functions of management and how they are performing in relation to them. This performance measurement is encapsulated within the Management Functional Assessment Model (MFAM).

Having outlined the seven functions of management, we must also acknowledge that traditional approaches to management are inadequate for keeping pace with changes in construction-related dynamic operational environments (Bounds et al. 1994).

This being the case, a new approach for considering organisations and their environmental interaction has been developed and this new philosophy is termed 'post-modernism'. As noted by Jackson and Carpenter (2000): "That it [post-modernism] has relevance to the understanding of [construction] organisational behaviour is not in doubt". Thus the importance of a post-modernist philosophy would need to be incorporated into any change process/improvement model, in order to improve on the effectiveness of managers and hence corporate and project quality management activities.

In the next section we shall explore post-modernism in a little more detail. It is important that construction companies have an understanding that this approach can positively impact upon corporate and project performance and hence the quality of their product/service provision.

Post-modernist philosophy

Successful construction companies are those that have changed their business processes due to re-evaluating their understanding of how business activities should be organised/conducted. They may have had to abandon previous organisational and operational procedures and created new, more appropriate ones. It is likely that previous procedures were based upon assumptions related to technology, people and organisational goals which may no longer hold true and hence lack validity within their changed operational environments.

Modernist theory assumes that change is a linear process and can be managed in an incremental way with distinctive points of conception and completion. In essence it is a belief in a simple cause-effect relationship. However, a more realistic view of construction operational environments rejects the notion of linearity. Post-modernist organisations realise that change develops in many directions and the world is best understood in terms of disorder and unpredictability. A post-modern philosophy recognises the need for versatility, and the emphasis is placed on organisational flexibility and quick response.

Within the following section a more detailed analysis of the differences has been undertaken and the advantages of post-modernism established for construction companies/managers.

Modernism versus post-modernism

In times of static or limited dynamic environmental change, a modernist organisational structure can cope with changes reasonably well. However, when the environment becomes more dynamic and complex, the modernist-structured organisation finds it difficult to cope with the implications of change.

Passmore (1994) noted: "Most of us are born with a good deal of flexibility; it's a helpful trait that allows our species to adapt to the wide range of habits and circumstances we encounter. But the process of growing up in a hierarchical [as are some construction companies] world teaches us to become inflexible".

Passmore is therefore advocating that people can inherently deal with change and that it is the bureaucratic systems they work within that stifles their intrinsic flexibility. Most modernistic construction firms would fall under this umbrella.

Early authors on this subject, such as Weber (1908), purported: "Modern business enterprises are structured as 'rational-legal' hierarchical and bureaucratic systems characterised by standardised operating procedures, regulations, performance standards and 'rational' decision-making processes [not suitable for deploying TQM/EFQM.E.M] that are based upon technical and professional expertise".

The above is now being contested by various authors such as Morris and Brandon, who suggest that there has been a paradigm shift in the way construction organisations view themselves and their operational environments. After all: "When the business world undergoes change, only those companies that react quickly will prosper. The ability to react requires considerable flexibility and openness to new ideas and approaches. In creating this foundation the basic assumptions of the business must be re-examined" (Morris and Brandon 1993).

The above noted paradigm shift is apparent in the post-modern organisation.

Structure of relationships

Within modernist organisations there exist very simple structure or boundary relationships. Linkages are achieved through formal rules and procedures, and relationships between different groups are formalised. In comparison, the post-modernist organisation possesses little distinctiveness of roles, and boundaries are blurred. The emphasis is placed upon creating teams with positive productive relationships, all directed at increasing the organisation's ability to cope with change, because this is necessary for them to be creative. Majaro (1992) points out that making the change to a post-modernist organisation "is easier said than done" and that "one of the most difficult challenges to any organisation is the process of changing a climate or corporate attitudes". It is undoubtedly a difficult change process for any construction organisation to undergo, but the benefits are well worth the effort.

Hierarchy

Modern organisations have a very distinct hierarchy, with clearly defined leadership roles fixed by legitimacy and tradition; there are leaders and followers. Contrasted with this is the post-modern organisation, where normal hierarchy does not exist and staff act according to agreed-upon areas of expertise. The term for this approach is 'hecterarchy', in which very high levels of fluidity are maintained. The high level of fluidity is a basic necessity for construction organisations because "[t]oo much is changing for anyone to be complacent" (Peters 1988). As construction organisations move to areas of increased complexity of service, there is a requirement to implement increasing hecterarchic ways of operating, and construction companies/managers must be cognisant of this fact.

Mechanistic versus holographic

In the modern construction organisation, the relationships between tasks are of a mechanistic nature and there is a high degree of linear relationship between tasks. However, within the post-modern organisation, high levels of group work exist, each with a correspondingly high level of autonomy. The overriding linking force binding these empowered groups together is that of organisational culture. This form more readily suits the reality of today's changing business environment because organisations, and the markets they operate in, are messy and not linear. Building a shared culture and conception of the world takes a great deal of time and effort on the part of construction firms. Traditionally some less enlightened construction companies have had a culture based upon mistrust and the use of frequent sanctions by managers and senior managers.

Determinacy versus interdeterminacy

The modernist construction organisation conducts all matters in a determinate manner, where a high degree of emphasis is placed on imposed stability, control and discipline. This assumes that a construction firm can exercise a high degree of control over its environment, but this is not the case in reality.

Post-modern organisations conduct matters in such a way that they emphasise indeterminacy. This is an acknowledgement that the environment is highly unpredictable and uncertain and it has different values from the modernistic organisation – for example, flexibility and innovation are highly prized traits.

> Flexible people are open minded, willing to take reasonable risks, self-confident, concerned and interested in learning. They are creative and willing to experiment with new behaviours in order to make better choices about what works for them and the organisation. They possess basic skills that allow them to adapt readily to new circumstances, and they view themselves as able to make the best of opportunities that come their way.
>
> *(Passmore 1994)*

This in essence is the post-modernistic organisation and one construction firms/managers should seek to emulate.

Causality

A major difference between modern and post-modern construction organisations is that modernistic ones view causality as having a linear relationship. They view every element of organisational life as having a cause and effect and consequently they manage their organisation in this light. However, post-modernistic ones think in terms of a 'circle'. They are encouraged to look for complexity and the interconnection of cause and effect. This demands a high level of staff participation and makes good management sense. The rationale for participation has been stated as follows: "When subordinates are consulted about and contribute to the change process [for example, quality improvement processes] many benefits accrue" (Sayles 1989).

A more enlightened view of corporate strategy is that it is in fact 'emergent' and not linear, and hence construction companies can better cope with this if they have adopted a post-modernist approach to business and people.

Morphostatic versus morphogenic

Morphostatic processes are defined as those that support or preserve the present mode of operation and include formal and informal control systems, with the emphasis being placed on procedures. This approach is not at all conducive to learning, or seeking to improve corporate/project performance. A more enlightened approach is adopted by the post-modern construction organisation, where a morphogenic culture exists. Morphogenic processes are those that tend to allow for change and development, and the exciting nature of change is always advocated and accepted. This type of construction organisation encourages staff to be proactive in all areas and functions, and hence always seeks to learn and improve corporate/project performance.

Customer Focus

Construction companies have to be open to ideas related to organisational improvement, thus they should be learning organisations (this issue is expanded upon in the next section of this chapter). They must consider new management tools and procurement systems that have proved to be most advantageous in manufacturing industries. These tools and methods of operation can assist in differentiating a company from its competitors. The application of a post-modernist approach to managing construction companies could provide the following advantages:

- Companies are more flexible and therefore better able to cope with the demands of a changing and challenging operational environment.
- Companies attain enhanced teamwork and participation at all levels of the company, leading to improved communications.
- Organisational culture is highly motivated and proactive, leading to increased participation and productivity.
- Enhanced corporate innovation and learning become embedded.
- Improved product/service quality becomes the norm.
- There is greater market awareness and thus enhanced stakeholder satisfaction.

The identified characteristics of the post-modernistic company are essential for a construction organisation to be able to operate both efficiently and effectively in a dynamic and turbulent operational/economic environment.

Construction firms require variety in their approach, and hierarchical authoritarian organisations are poorly equipped to provide such variety. Only construction organisations based on the post-modern model, with vastly reduced bureaucratic control, a rich array of horizontal communication channels and personnel who are given a substantial share of authority to make choices and to develop new ideas, are likely to survive.

With the above in mind, the MFAM has been developed. Due to its requirement for continually evaluating the management functions and activities, it truly enables a move towards a post-modernist philosophy to be achieved, set within a learning organisational culture.

For example, as indicated in *Table 5.1*, under criterion 2, Organising, can be found the sub-criteria of:

2.1 Creating the correct organisational structure.
2.3 Creating a self-learning organisational culture.
2.4 Developing a value system based upon enhancing performance.

These are not mutually exclusive from the requirements of post-modernism (see *Table 5.1* for further elements of post-modernism encapsulated in the MFAM).

TABLE 5.1 Marking criteria for MFAM

Criteria	Assessment Category	Max	Score
1. Forecasting/ planning	1.1 Setting the objective and strategic planning process in motion	4	
	1.2 Gathering and analysing information related to clients and markets	4	
	1.3 Detailing business processes	4	
	1.4 Gathering and analysing information related to competitors and benchmarking	4	
	1.5 Resources planning	4	
	Criterion total	20	
2. Organising	2.1 Creating the correct organisational structure	4	
	2.2 Establishing appropriate authority and responsibility for all personnel	4	
	2.3 Creating a self-learning organisational culture	4	
	2.4 Developing a value system based on enhancing performance	4	
	2.5 Deployment of new technology linked to corporate enhancement	4	
	Criterion total	20	

(Continued)

TABLE 5.1 (Continued)

Criteria	Assessment Category	Max	Score
3. Motivating	3.1 Developing a co-operative culture based upon stakeholder satisfaction	4	
	3.2 Ensuring staff have the skills, competences and resources to perform set tasks	4	
	3.3 A consideration of personnel needs linked to self-actualisation	4	
	3.4 Engagement in processes, increase areas of responsibility and self-monitoring	4	
	3.5 Results satisfaction – feedback on performance in a timely manner	4	
	Criterion total	20	
4. Controlling	4.1 A monitoring system for each key stage of business process	4	
	4.2 Measuring performance levels	4	
	4.3 Determining customer satisfaction levels	4	
	4.4 Determining the efficiency and effectiveness of resource utilisation	4	
	4.5 Conducting a comparative analysis between set targets and actual results, leading to appropriate actions	4	
	Criterion total	20	
5. Co-ordinating	5.1 Unity of all other functions	4	
	5.2 Establishing effective internal communications	4	
	5.3 Developing a conflict solving culture	4	
	5.4 Updating deviations: revision and possible re-coordination of other resources	4	
	5.5 Ensure effective information management	4	
	Criterion total	20	

Definition of a learning organisation

There is no clear consensus as to what constitutes a learning organisation, and a multitude of definitions abound. These range from aspirational type definitions of organisations,

> where people continually expand their capacity to create the results they truly desire, where new and expansive patterns of thinking are nurtured, where collective aspiration is set free, [truly in line with a post-modernist philosophy] and where people are continually learning how to learn together
>
> *(Senge 1990)*

to more normative definitions such as that espoused by Garvin,

> A learning [construction] organisation is an organisation skilled at creating, acquiring, and transferring knowledge, and at modifying its behaviour to reflect new knowledge and insights.
>
> *(Garvin 1993)*

Garvin's views have been fully encapsulated within the MFAM.

Nyhan (2004) suggested that "the prescriptive and simplistic formula based view of the learning organisation does nothing more than discredit the concept". In their opinion, becoming a learning organisation involves more than simply applying a formula; each individual organisation needs to "devise its own unique theory based on its own distinctive practice" (Nyhan 2004). Once again the advocated MFAM empowers this approach.

Historical development

The concept of the learning organisation has been around for quite some time; Burns and Stalker (1961) published their theory of mechanistic and organic systems following lengthy studies of a large number of Scottish electronics companies operating in increasingly competitive and innovative technological markets during the 1950s.

The 1980s was a decade of immense upheaval for many large corporations which increasingly found "their success eroded or destroyed by the tides of technological, demographic, and regulatory change and order of magnitude productivity and quality gains made by non-traditional competitors" (Hamel and Prahalad 1994).

A new wave of literature appertaining to learning organisations emerged during this period, heavily influenced by organisational learning and action learning theories such as those developed by Revans (1983). Much of the work from this period recognised that corporate survival in the new global competitive environment was dependent on an organisation's ability to learn faster than its competitors, and that this ability may be the organisation's only form of sustainable competitive advantage (de Geus 1988).

Nonaka (1991) recognised that in a global economy typified by shifting markets and technological proliferation, successful construction companies will have to "consistently create new knowledge, disseminate it widely throughout the organisation, and quickly embody it in new technologies and products".

However, it was Senge's seminal text, *The Fifth Discipline* (Senge 1990) that really popularised the concept of the learning organisation. Senge described five vital dimensions or 'disciplines' that he considered to be essential for an organisation to become a truly learning company; they are team learning, personal mastery, managing mental models, shared vision and systems thinking.

The fifth discipline, systems thinking, was seen as the integrating discipline that unites the organisation, individual and total environment, based on a conceptual framework that describes a system as a set of interrelated subsystems. Senge (1990) proposes that it is the relationship between these subsystems that ultimately influences the functioning of the whole. This concept is not mutually exclusive from the concept of the integrative nature of the seven functions of management and their impact upon construction organisational performance at corporate and project level.

Senge's work, however, has been criticised for paying insufficient attention to knowledge management systems, the structures of the organisation and their implication as a resource to learning (Sun and Scott 2003), whilst Garvin (1993) considers Senge's model too ethereal and lacking a 'framework for action'.

Culture and the learning organisation

Much of the discussion in the management literature is clearly written from the perspective that the learning organisation can be designed and managed effectively to produce positive outcomes for the organisation. Many commentators have attempted to specify what the culture of a learning organisation should consist of. Although numerous authors (e.g. Garvin 1993; Senge 1990) have considered the notion of a learning organisation culture, there is no widely accepted theory or view on this issue. Some have identified specific attributes of a learning organisation culture such as entrepreneurship and risk taking. Indeed the literature on learning culture characteristics is extremely broad, drawing on work from sociology, psychology and anthropology as well as business disciplines, which perhaps makes the task of formulating such a theory a monumental one.

Organisations and learning

Pedlar and Aspinwall (1998) point out that the aforementioned writings (Senge 1990) take such a wide view of the structures which an organisation does or needs to learn that the idea of learning becomes lost. They cite four questions which are of relevance here: What is learning? Are there types or levels of learning in organisations and are they recognised? What are the different levels of learning? How does an organisation facilitate or inhibit the learning process? Construction companies must fully consider these important questions.

They suggest that there is a need to hold on to the idea of the learning organisation as a direction, while organisational learning, which is a fundamental component of the learning organisation, is seen as a heuristic device to explain or quantify learning activities. This would seem to suggest that an emphasis be put on understanding how learning is defined, acquired and used at the individual and organisational level.

Rarely do construction firms have an understanding of what it is they are measuring and, when they do, they may be only measuring activities as part of an organisational control system. However, a major challenge for them will be to develop valid measures of learning outcomes specifically to assess whether they have actually learned, demonstrated by changed behaviour and project/corporate improvement.

Learning cultures

The concept of culture itself is intangible and the notion of a learning culture is perhaps easier to experience than describe. There is evidence, however, to suggest that an organisation's culture may facilitate or inhibit learning depending on its characteristics (Argyris and Schön 1978). They suggest that an organisation's defence routines may be both anti-learning and over-protective. They further argue that such patterns of behaviour may become so embedded in the culture that they are rarely questioned or challenged.

Cummings (2005) emphasises that it is important for the firm's culture to be supportive because it is difficult to develop and sustain appropriate learning behaviours if the corresponding organisational values are not in place (requires a supportive paradigm, e.g. morphogenic), and similarly values are difficult to sustain if the appropriate incentives and examples do not exist. This suggests a synergistic relationship between the elements of culture and learning activities within the organisation systems which support the learning organisation. He further suggests that culture often embodies an accumulation of prior learning, based on earlier success.

Underpinning cultural values

Building construction learning organisations is, in effect, an attempt to manage the culture of the said organisation and it requires specific attention to some key cultural values if it is to be a successful undertaking. For example they need to address:

- *Celebration of success.* If excellence is to be pursued with vigour and commitment, its attainment must be valued within the organisational culture.
- *Absence of complacency.* Learning organisations reject the adage 'if it isn't broke don't fix it'; they are searching constantly for new ways of delivering products and services. Thus innovation and change are valued and promoted within the organisation.
- *Tolerance of mistakes.* Learning from failure is a prerequisite for progressive organisations. This in turn requires a culture that accepts the positive spin-offs from errors, rather than one that just seeks to allocate blame. However, this does not imply a tolerance of routinely poor or mediocre performance from which no lessons are learned.
- *Belief in human potential.* It is people that drive success in organisations, using their creativity, energy, and innovation. Therefore, the culture within a learning organisation values people and fosters their professional and personal development.
- *Recognition of tacit knowledge.* Learning organisations recognise that those individuals closest to processes have the best and most intimate knowledge of their potential and flaws. Therefore, the learning culture values tacit knowledge and shows a belief in empowerment (the systematic enlargement of discretion, responsibility, and competence).
- *Openness.* Because learning organisations try to foster a systems view, sharing knowledge throughout the organisation is one key to developing learning capacity. 'Knowledge mobility' emphasises informal channels and personal contacts over written reporting procedures. Cross-disciplinary and multifunction teams, staff rotations, on-site inspections and experiential learning are essential components of this informal exchange (a post-modern approach).
- *Trust.* For individuals to give of their best, take risks and develop their competencies, there must be demonstrable trust.

Örtenbland's (2004) 'learning structure' model, builds on this idea by describing a decentralised, flatter organisational structure that is team based, with learning depicted as an input and flexibility as an output.

Providing the correct corporate environment

Garvin (1993) suggests that a learning organisation is one that fosters "an environment that is conducive to learning". He purports that in order for employees to learn, they need "time for reflection and analysis, to think about strategic plans, dissect customer needs, assess current work systems, and invent new products". This highlights an important prerequisite for implementing any new initiative: the provision of adequate resources, particularly those of time and funding, which are not mutually exclusive.

However, time and money alone will not create the required climate for learning. Ho (1999) proposes that the learning organisation provides an environment where "people are excited in trying out new ideas and recognise that failure is an important part of success".

Love (2004) underpins this view, describing an atmosphere where *"experimenting with new approaches is encouraged [not common in construction] and errors are not perceived as failures"*. These traits, when viewed in the context of an organisational environment typified by ever-increasing complexity and uncertainty (Malhotra 1996), clearly point towards a requirement for a morphogenic culture utilising processes that "allow for change and development . . . [and where] the exciting nature of change is always highlighted" (Griffith and Watson 2004).

Knowledge management

In the late 1980s, Pedler et al. (1988) recognised the importance of utilising information technology to "informate as well as automate . . . [in order to] seek information for individual and collective learning". However, Lobermans has asserted that a "corporate architecture" needs to be in place to facilitate learning and to "create knowledge sharing and dissemination mechanisms across the organisation" and that the capture and systemisation of knowledge is a prerequisite to being a learning organisation (Lobermans 2002). The growing number of organisations utilising intranets and 'lessons learned' databases gives some indication of the perceived value of knowledge management systems to the construction industry.

However, recent research into cross-project learning led Newell to conclude that "there is accumulating evidence that the medium of capture and transfer through ICT [information and communications technologies] such as databases and corporate intranets is limited in terms of how far such technology can actually facilitate knowledge sharing" (Newell 2004). Newell's study also found that where transfer of learning had occurred, it had depended far more on social networks and a process of dialogue than on ICT.

These findings concur with the view of Nonaka (1991) in that the key to construction organisations gleaning greater knowledge is through facilitating:

- the sharing of tacit knowledge through socialisation;
- the collation of discrete pieces of explicit knowledge to create new knowledge;
- the conversion of tacit knowledge into explicit knowledge, i.e. externalising what individuals know; and
- the conversion of explicit knowledge to tacit knowledge, i.e. internalising explicit knowledge.

The key features of construction learning organisations relate less to the ways in which organisations are structured and more to the ways in which people within the organisation think about the nature of, and the relationships between, the outside world, their colleagues and themselves.

Of course, the key focus for all organisational activities should be the satisfaction of the client, and the learning culture is directed to this end product.

Crucially, learning construction organisations do not focus exclusively on correcting problems or even on acquiring new knowledge, understanding or skills. They aim instead for more fundamental shifts in organisational paradigms and try to encourage the development of learning capacity at all levels.

Mental models

Senge's (1990) discipline of managing mental models recognises that "new insights fail to get put into practice because they conflict with deeply held internal images of how the world works, images that limit us to familiar ways of thinking and acting".

Argyris and Schön (1974) opined that people are often unaware that the mental models that inform their actions are often not founded in the beliefs that they explicitly advocate, leading to a contradiction between their espoused theory and their theory in practice. In order for people to manage their behaviour more effectively, they suggest the use of double loop and even triple-loop learning, in order to develop congruence between theory and deployed practice.

The double-loop learning advocated by Argyris and Schön is fundamentally what Senge was referring to when he suggested that mental models should be brought to the surface and reflected on by "balancing advocacy and inquiry", a process he describes as being "open to disconfirming data as well as confirming data – because we are genuinely interested in finding flaws in our views" (Senge 1990).

This contemplative approach is necessary in order for construction organisations to escape what Shukla calls "the success trap" (1997). He describes how successful construction companies try to replicate their achievements by formalising their effective practices and procedures, standardising their products and services and investing in tried and tested technologies.

This single loop approach to learning results in construction firms becoming less sensitive to competitive demands; they lose touch with their environment and, as Shukla explains, "their past learning becomes a hindrance in the way of the necessity of new learning; they must 'unlearn' to learn" (Shukla 1997).

Hamel and Prahalad (1994) use the term 'frame' in place of 'mental model', proposing that "[a]lthough each individual in a [construction] company may see the world somewhat differently, managerial frames within an organisation are typically more alike than different" and "[a]lmost by definition, in any large organisation there is a dominant managerial frame that defines the corporate canon".

The suggestion that there can be an institutional model echoes the view espoused by de Geus, who sees the mental model of each learner as "a building block of the institutional mental model" (de Geus 1988) (Cummings 2005).

Single, double and triple-loop learning explained in more detail

Argyris and Schön (1974) first developed the idea that there are two basic types of organisational learning, 'single loop' or 'double loop'. Single loop, as noted, is the type of learning where organisations respond to changes in their internal and external environments by detecting and correcting errors in order to "maintain the central features of the organisational norms" (Barlow and Jashapara 1998). Argyris (1996 cited in Dahlgaard 2004), when considering learning within an organisational context, suggests that an error is any mismatch between the intention and what actually happens (the results). However, he further argues that discovering errors is not really learning and that learning only occurs when the discovery or insight is followed by an action. From this viewpoint, learning inevitably involves the taking of some action.

Single-loop learning

Single-loop learning assumes that problems and their solutions are close to each other in time and space (although they often aren't). In this form of learning, we are primarily considering our actions. Small changes are made to specific practices or behaviours based on what has or has not worked in the past. This involves doing things better without necessarily examining or challenging our underlying beliefs and assumptions. The goal is improvements and fixes that often take the form of procedures or rules. Single-loop learning leads to making minor fixes or adjustments.

It could be argued that incremental, imitative learning methods such as benchmarking and best practice are examples of single-loop learning. Within what Argyris described as 'single-loop' learning, decisions are based solely on observations, while in double-loop learning decisions are based on both observation and thinking.

Learning hasn't really taken place until it's reflected in changed behaviours, skills and attitudes (Stata 1989).

Double-loop learning

Double-loop learning involves a more demanding approach to learning, where an organisation's norms, policies, assumptions and past actions are critically examined in order to inform new strategies for learning (Argyris and Schön 1974). Inevitably, such introspective organisational analysis may bring about conflict. Love (2004) maintains: "Frequently organisational conflict is a correlate of double loop learning in as much as the status quo is challenged [as it moves towards a more morphogenic culture]".

In summary it can be stated that in 'single-loop learning', people's decisions are based solely upon observations, while in 'double-loop learning', decisions are based on both observation and thinking.

Double-loop learning leads to insights about why a solution works. In this form of learning, we are primarily considering our actions in the framework of our operating assumptions. This is the level of process analysis where people become observers of themselves, asking, "What is going on here? What are the patterns?" And we require this insight in order to understand the pattern. Double-loop learning works with major fixes or changes, like redesigning an organisational function or structure.

In 'triple-loop learning' a reflection phase is incorporated to support or improve the thinking phase and hence to improve the decision making process. "Thus both double and triple loop learning can be considered as generative learning, while single loop learning can be considered an adaptive learning" (Dahlgaard 2004).

The developed MFAM employs the concept of triple-loop learning.

Triple-loop learning

Triple-loop goes beyond insight and patterns, and the result creates a shift in understanding the corporate context, or point of view, where new commitments and ways of learning are produced. This form of learning challenges us to understand how problems and solutions are related, even when separated widely by time and space. It also challenges firms to understand how previous actions created the conditions that have led to their current problems. The relationship between organisational structure and behaviour is fundamentally changed because the organisation learns how to learn.

Summary

The concept of the learning organisation has evolved as a response to a rapidly changing, dynamic business environment which is constantly in flux. The idea, then, of a fluid, flatter, less hierarchical organisational structure that offers less resistance to the seepage of knowledge through the organisation appears to have credence.

An organisational structure provides only the skeleton of the learning organisation; a capillary system is necessary in order to transfer knowledge around the organisation at all levels. It does seem that most knowledge management strategies focus solely on the electronic collation of information, failing to take account of how different types of knowledge are internalised and externalised via the use of already existing social networks.

There also appears to be a degree of consensus that a 'learning climate' has to be created, where individuals feel free to experiment with new ways of doing things. However, this requires a blame-free culture where mistakes, instead of being hidden, are acknowledged and learned from. Changing organisational culture requires a well-planned change management strategy to be developed, and this has to be initiated and supported by senior management for it to have any chance of success. It does seem that the utilisation of 'mental models' by construction companies inhibits the implementation of new concepts, and most models are based on replicating previously effective practices. The models, though individually held, collectively form and reinforce the organisational model, which is focused on maintaining the status quo. The MFAM is designed to challenge the status quo, with a view to obtaining organisational improvement.

The idea of surfacing mental models (Senge 1990) seems closely aligned with the concept of double-loop learning (Argyris and Schön 1978). The introspective organisational analysis associated with both concepts is a quantum leap away from the morphostatic culture (Griffith and Watson 2004) prevalent in many organisations, and may prove to be one of the most difficult learning organisation characteristics to attain. There is clearly a need to make a change management strategy an integral part of any generic implementational model.

Other key characteristics that typify a learning organisation are:

- a strategy for creating, acquiring and disseminating knowledge;
- collective aspiration (a shared vision);
- an emphasis on continuous learning leading to continuous improvement;
- a holistic, 'systems thinking' approach to learning that recognises the interrelatedness of the organisation, the individual and its external environments; and
- a tolerance of some experimentation by people.

Several problematic issues may prevent a construction company from successfully implementing learning organisation concepts. For example, organisational structures geared towards stability rather than change as identified by Johnson and Scholes (2002) are noted as an unsuitable framework upon which to found aspirations to become a learning organisation.

A lack of senior management support, resulting in failure to provide adequate resources, particularly in respect to allowing employees 'time to think', will also lead to failure.

The above has been highlighted in order to emphasise the importance of triple-loop learning being incorporated into any model designed to improve the effectiveness of management functions. The MFAM does indeed incorporate 'triple-loop learning'.

Management Functional Assessment Model (MFAM)

The behaviour of an organisation's leaders should create a clarity and unity of purpose within the company and an environment in which its personnel can learn and improve. A truly empowered organisation employs both a top-down and bottom-up approach to managing and performing its organisational activities.

Construction companies perform more effectively and efficiently when all interrelated activities are understood and systematically managed, and decisions concerning current operations and planned improvements are made using reliable information that includes both stakeholder perceptions and expectations.

Corporate performance is maximised when it is based on the management and sharing of knowledge within a culture of continuous triple-loop learning, innovation and improvement. The penalties of failure must not outweigh the rewards of success, or this will undermine any attempt to encourage a culture of innovation and risk taking set within the context of a learning organisation culture.

A construction company works more effectively and efficiently when it has mutually beneficial relationships built on trust and the sharing of knowledge and integration

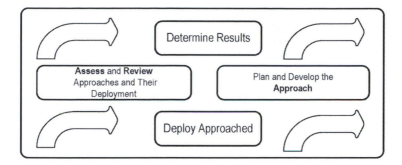

FIGURE 5.1   The criteria underpinning the RADAR concept

with its partners. Therefore leadership and culture are vital components of a continuous improvement learning process. This is recognised by the European Foundation for Quality Management in the development of its RADAR concept. RADAR is indicated in *Figure 5.1* (European Foundation for Quality Management 1999).

Obtaining a sustainable competitive advantage via MFAM deployment

For a construction company to attain a sustainable competitive advantage, it requires a competitive-orientated management system. The advocated management system should embrace the issues previously described within this chapter, namely the seven functions of management, a morphogenic philosophy and performing as a learning organisation utilising triple-loop learning, all as part of a holistic management process.

The system should also address key concepts such as leadership, personnel and development. However, the system must also fully address the needs of a company's stakeholders.

The MFAM is based upon previous works including those of Deming, Baldrige and the European Foundation for Quality Management. However, it is original and when deployed provides an effective link between all organisational activities set within a framework of corporate profitability and stakeholder satisfaction.

This section explores the model's constituent parts and relates them to the process of gathering data on organisational performance and requirements, focused upon attaining/maintaining a competitive advantage. The key concept of 'RADAR', which forms part of the European Foundation for Quality Management Excellence Model, is embedded within the MFAM, thus embedding the triple-loop learning concept.

Corporate excellence is measured by an organisation's ability to both achieve and sustain a competitive advantage through satisfying its stakeholders. This can only be achieved by the efficient and effective utilisation of all corporate resources, which include the 5M's. The 5M's have to be treated as holistic, and the MFAM provides a means for setting corporate objectives that are linked with stakeholder expectations and needs. The advocated model enables construction firms to monitor and benchmark their activities and further enables them to score their performance in key operational areas in a way that leads to enhanced project and corporate performance.

By adopting a holistic approach to customer requirements, building on stakeholder contribution, not only can 'added value' be attained, but it can also be measured. Once measured, a benchmark can be set for engaging in a continued drive for organisational excellence.

The MFAM addresses the critical issue opined by Hersey and Blanchard (1972 cited in Hutchin 2001). "How do managers cope with the inevitable barrage of changes, which confront them daily in attempting to keep their organisations viable and current? While change is a fact of life, effective managers . . . can no longer be content to let change occur as it will, they must be able to develop strategies to plan, direct and control change".

The MFAM provides a means for managers to address the above key question by providing a structured approach to change management focused upon an organisation's (and its managers') activities in a drive for corporate excellence.

The incorporation of triple-loop learning enables construction managers to be pro-active in relation to change, and this could prove to be most beneficial for all stakeholders.

Developing measurement tools

Given the current attention to becoming learning construction organisations, it seems appropriate that we begin to formalise a measurement method and communications model that will further enhance learning, reinforce positive outcomes and minimise negative outcomes for construction companies.

The model presented is a 'functional assessment model'. However, the functional assessment model forms part of 'competitive-orientated management'. This is a system of management designed to gain and sustain a competitive corporate advantage. The concept of competitive-orientated management may be represented as a tetrahedron as depicted in *Figure 5.2*. It is based upon the principles of competitive

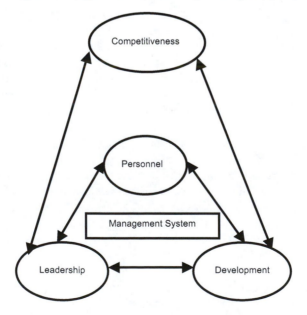

FIGURE 5.2 Competitive-orientated management core concept

achievement, leadership, personnel and development. Hardy (1983) states that the development of a competitive advantage automatically creates an opportunity, and so the reasoning may be modified so that successful businesses are engaged in the creation and exploitation of competitive advantages.

Constituent parts of the tetrahedron

According to Day and Wensley (1988) the essence of competitive advantage is the conversion of superior skills and resources into positional advantages, which in turn create positive outcomes. A competitive advantage is sustained only if it continues to exist after efforts to duplicate that advantage have ceased. Watson and Chileshe (1998) explored the linkages between TQM and competitive advantage and found that organisations implementing TQM had improved their efficiency and effectiveness. Other sources of competitive advantage may be obtained through the use of benchmarking best practice (Shakantu and Talukhaba 2002), organisational learning or organisational strategic alliances (Ngowi 2001). The following subsection now examines and discusses the constituent parts of the tetrahedron.

Leadership

The role of senior management is critical to the success of any change process, and effective leadership has to be demonstrated during the change process. All employees have to be given the time and skills to make a useful contribution towards a drive for a sustainable competitive advantage. "One factor that affects the individual's reaction to change is their past experiences. Individuals also resist having solutions imposed on them. [I]t is therefore necessary [for construction managers] to obtain the commitment of the individuals to the change initiative" (Kotter 1999 cited in Martin & Cullen 2005).

Personnel

Employees have to be motivated to engage in a corporate culture directed at achieving stakeholder satisfaction. Senior management must not forget that employees are also internal stakeholders of the corporate entity. Thus senior managers should remember to engage in 'consultation' before 'implementation' when engaged in setting and deploying corporate plans. This aspect allows staff to make a valid contribution to the decision-making process. Culture is a vital aspect for consideration and has already been expanded upon in this chapter under the seven functions of management.

Development

A morphogenic culture should be the aim of senior management; the critical issue of staff motivation has to be addressed and this aspect has also been covered under the seven functions of management. The development of a construction learning organisation requires the consideration of both business processes and environmental issues/aspects. Therefore environmental scanning has to be deployed in order to establish relevant external influences, and this may be done by conducting a Strengths, Weaknesses, Opportunities and Threats (SWOT) analysis and/or a Political, Economic, Social, Technological, Environmental and Legal (PESTEL) analysis. These processes should be linked to the RADAR concept; also, full engagement with the concept of a learning organisation is a fundamental requirement for organisational development.

Rationale for MFA model – development and deployment

Excellence models are strongly related to quality and quality prizes/awards, e.g. the Deming prize in Japan, the Malcolm Baldridge National Quality Award (MBNQA) in the USA and the EFQM excellence award in Europe.

Some of these models have been in existence for decades, yet interest in them has increased. Construction organisations do not just engage with the models in order to win a prize, they are used in order to guide a firm in achieving organisational excellence. They represent a coherent approach to organisational 'management policies' and help focus a company's attention on critical analytical assessment criteria (Goasdove 2001 cited in Hermel & Ramis-Pujol 2003). As noted by Hermel and Ramis-Pujol (2003), "The above models have quickly entered the management practices roller-coaster".

Various critics have pointed out that some companies that have previously won prizes have in the long term not performed favourably. For the most part, the real reason for these failures is poor-quality management and inappropriate strategies that are not easily related to by the managers who are responsible for implementing them (Heller 1997 cited in Hermel & Ramis-Pujol 2003).

Beechner and Hamilton (1999) comment that many failures (noting the above) may be attributed to a lack of any attempt at integration and the misalignment of strategic planning, continuous improvement and the transfer of knowledge when trying to deploy excellence models (cited in Hermel & Ramis-Pujol 2003). The MFAM seeks to address these noted critical issues.

Many construction managers do seem to have great difficulty in understanding concepts, and this could be a contributing factor to existing excellence model failure. In today's dynamic and very challenging economic operational environments, implementation is viewed not as a choice between options but instead as the "art of balancing among those options". What to balance are aspects that could be further investigated (Pascale 1992).

The complexity of the excellence model's suitability and implementation is further complicated when one adds to it the size of a construction company attempting to deploy such a model. The MFA model provides the focus that managers require in order to be proactive in the management of change processes. At the same time it also establishes a focal point for linking in a truly holistic manner the sometimes disparate functions of:

- setting and implementing strategic plans;
- setting and implementing operational plans;

- providing due consideration to organisational size when selecting and engaging in self-assessment linked to an improvement model;
- linking the various functions of management in an effective and efficient way;
- obtaining feedback for stakeholders on organisational performance, with a view to the enhancement of service and product provision; and
- building on the concept of triple-loop learning leading to continuous corporate improvement and enhanced customer satisfaction.

Thus the MFA model does address the issues noted above by Hermel and Ramis-Pujol (2003) and Pascale (1992). The MFAM provides a means of self-assessment for construction organisations related to the seven functions of management. Construction companies can utilise the model in order to establish how efficient and effective they are operating, and further, they can identify areas where they need to improve their project and corporate performance levels.

Management Functional Assessment Model (MFAM) constituent parts

The MFAM is based upon six functions of management (forecasting and planning being linked together under one heading, thus all previously noted seven functions are incorporated). These are forecasting and planning, organising, motivating, controlling, co-ordinating and communicating. The first five functions are encapsulated within a framework of an effective and efficient system of communication (see *Figure 5.3*). The MFAM has been designed to aid managers in determining

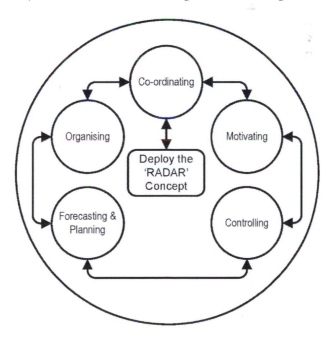

FIGURE 5.3   Management Functional Assessment (MFA) incorporating the RADAR concept

the key activities to be addressed in order to improve corporate efficiency and effectiveness.

Communicating

Figure 5.3 indicates that communication encompasses the other functions, because construction managers engaging in communication are active in a two-way process; they also will be on the receiving end of communications (Fryer 1997).

Forecasting and planning

This criterion is concerned with the shape of future strategy both short and long term; its function is to answer three basic questions:

*   Where is the company now in terms of its vision and mission?
*   Where does it want to be as part of its future corporate plans?
*   How is the company to achieve its set aims based upon forecasts?

The forecasting and planning criterion contains five basic categories:

1.1 Setting the objectives and strategic planning process in motion.
1.2 Gathering and analysing information related to both clients and the markets (all stakeholders).
1.3 Detailing business processes (who, what, when, where and why).
1.4 Gathering and analysing information relating to competitors and benchmarking.
1.5 Planning resources (incorporating the 5M's).

Organising

The main managerial tasks here are to organise business processes with a concentration on maximising effectiveness and efficiency. The organising criterion contains five basic categories:

2.1 Creating an appropriate organisational structure.
2.2 Establishing appropriate authority and responsibility for all personnel.
2.3 Creating a self-learning organisational culture (morphogenic/post-modern).
2.4 Developing a corporate value system based upon enhancing performance.
2.5 Deploying new technology linked to corporate enhancement and competitive advantage.

Motivating

It has to be noted that motivation has many aspects both intrinsic and extrinsic, as established when considering the seven functions of management earlier in this chapter. The motivation criterion contains five basic categories:

3.1 Developing a co-operative culture based on stakeholder satisfaction.
3.2 Ensuring that staff have the skills and competencies to perform set tasks.
3.3 A consideration of personnel needs linked to self-actualisation.
3.4 Involvement in processes, increased areas of responsibility and self-monitoring (empowerment).
3.5 Results satisfaction – feedback on performance in a timely manner linked to RADAR.

Controlling

Control is dependent upon constant feedback from each stage of business processes, checking against quality specifications and measuring against performance indicators. A correct monitoring system allows for an increase in the efficiency and effectiveness of organisational activity. Organisations must consider feed forward of information for effective control, and this can only be fully achieved by deploying RADAR, which incorporates triple-loop learning. The controlling criterion contains five basic categories:

4.1 A monitoring system for each key stage of business processes.
4.2 Measuring performance levels (with an internal and external perspective).
4.3 Determining customer satisfaction levels.
4.4 Determining the efficiency and effectiveness of resource utilisation linked to project and corporate aims.
4.5 Conducting a comparative analysis between set targets and actual results, leading to appropriate and timely actions (RADAR).

Co-ordinating

The analysis of deviations of business processes and updating of the current plans in a holistic manner based on feedback and feed forward is a critical point in co-ordination. Again this can only be fully attained by the application of RADAR. The co-ordinating criterion contains five basic categories:

5.1 Unity of all the other functions.
5.2 Establishing effective communications.
5.3 Developing a conflict-solving culture, linked to corporate enhancement.
5.4 Updating of deviations: revision and the possible re-coordination of resources.
5.5 Information management – information has to be timely and in sufficient detail to inform corrective actions (RADAR).

Communicating

This is the link and the life-blood of corporate activity, and its effectiveness is measured within the context of the other five functions. Respondents are required to complete a score card based on organisational performance. A scoring of 0–4 is given in accordance with the criteria shown in *Table 5.2.*

TABLE 5.2 Scoring criteria to be applied to the MFAM

Score	Criteria for Scoring
0	No activity has been demonstrated.
1	Little activity has been demonstrated in this area.
2	Activity utilised but its use is sometimes dependent upon the situation, not a consistent approach.
3	The activity is deployed permanently and systematically.
4	The activity is deployed permanently and systematically, monitored and reviewed via benchmarking for improvement purposes; triple-loop learning is employed via RADAR.

TABLE 5.3 Summary of results for scoring the MFAM

Maturity Level	Total Score Allocated	Assessment Results
I	(0–20)	No methodology or clearly developed processes have been demonstrated. Management's purposes and functions are not clearly defined. For further development, it is necessary to reconsider the basic systems and corporate core business principles.
II	(21–40)	A methodology and some processes are in evidence but are erratic in their application. Managers should further develop their leadership skills, define organisational purposes more clearly and develop a strategy based on sound TQM/EFQM.E.M principles.
III	(41–60)	Management systems and processes are in evidence and utilised, and their approach is evaluated. It is necessary to pay attention to the optimisation of business processes and the improvement of quality at each stage. Need to perfect a control system linked to the importance of stakeholders.
IV	(61–80)	Management systems and processes are in evidence and linked to deployed approaches with constant quality checks within the management system taking place. Utilisation of external benchmarking in order to improve corporate performance has been demonstrated.
V	(81–100)	There is a clear demonstration of a systematic approach to strategy and policy setting linked to the deployment/approach undertaken. The management system is fully functional and the system is benchmarked and monitored in a drive for continuous improvement and the concept of RADAR is fully embedded.

Table 5.3 provides a summary of results for scoring the MFAM; the total management scoring process assists in determining the level of management maturity and its associated effectiveness.

MFAM analysis communication

The presentation of the analysis can be easily communicated to all staff via the application of a pentagonal profile. The scores can be plotted upon the profile and a corporate profile established. This process will also demonstrate where a construction organisation should place its efforts in order to improve their performance.

One must remember that action taken in one area will impact upon others. In other words the criteria are not mutually exclusive. Each time the MFAM is implemented and corrective actions taken, a new profile can be developed in order to benchmark the effectiveness of actions taken to improve corporate performance and hence address the competitive advantage issue. This activity becomes part of the RADAR approach.

The advantage of this approach is in the implementation of benchmarking and feeding forward the results and learning from them. Thus each time the MFA model is deployed, it is set within the context of RADAR and hence internal and external benchmarking is inherent. RADAR is explained in Chapter 4 of this text.

Linking RADAR and the Management Functional Assessment Model (MFAM)

The Management Functional Assessment Model incorporating RADAR encapsulates the facility for construction organisations to fully engage in a drive for continuous learning and improvement. Every time the MFAM is implemented and the scoring process applied, RADAR is embodied within the model. In this way forecasts and plans linked to deployment strategies are evaluated and appropriate actions determined via assessment and review. Only by employing this approach can the full benefits of MFAM deployment be attained by construction firms.

Figure 5.4 illustrates the critical linkages between the RADAR concept and the MFAM.

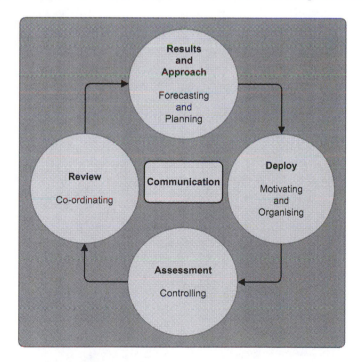

FIGURE 5.4   MFAM linked to RADAR

TABLE 5.4 Deployment of MFAM issues and advantages

Key Development Issues	Resulting Benefits
Process Improvements	A clear understanding of how to deliver value to clients and hence gain a sustainable competitive advantage via operations.
Attaining a construction organisation's objectives	Enabling the mission and vision statements to be accomplished by building on the strengths of the company and avoiding any weaknesses.
Benchmarking Key Performance Indicators	Ability to gauge what the construction organisation is achieving in relation to its planned performance targets.
Development of clear, concise action plans resulting in a focused policy and strategy	Clarity and unity of purpose so that the organisation's personnel can excel, learn and continuously improve.
Integration of improvement initiatives into normal operational activities	Interrelated activities are systematically managed with a holistic approach to decision making.
Development of group/team dynamics	People development and involvement. Shared values and a culture of trust, thus encouraging empowerment in line with post-modernist/morphogenic company culture.
Embed RADAR	Triple-loop learning and the self-perpetuating culture of learning and continuous project and corporate improvement.

The MFAM recognises that sustainable excellence in all aspects of performance is based on the management functions of forecasting/planning, organising, co-ordinating, motivating, controlling and communication. The application of the MFAM serves to address the key development issues shown in *Table 5.4* and empower the resulting benefits.

Excellence is dependent upon balancing and satisfying the needs of all relevant stakeholders and this includes people employed, customers, suppliers and society in general, as well as those within the organisation. The customer is the final judge of product and service quality and customer loyalty, retention and market share are best optimised through a clear focus on the needs of current and potential stakeholders/ clients.

Adopting an ethical approach and exceeding the society expectations and regulatory requirements best serve the long-term interests of any construction organisation. Corporate excellence is measured by an organisation's ability to both achieve and sustain outstanding results for its stakeholders, thus MFAM linked to RADAR with triple-loop learning has been developed and advocated.

The MFAM has been designed to aid construction managers in determining the key activities to be addressed in order to improve project and corporate efficiency and effectiveness within a framework for obtaining stakeholder satisfaction (Watson, Chileshe & Maslow 2005; Watson, Maslow & Chileshe 2005).

Conclusions

According to Harvey and Denton 1999 (cited by Hermel and Ramis-Pujol 2003), six fundamental developments provide the rationale for the importance and popularity of organisational learning, and they are:

- the shifting importance of the factors of production;
- the accelerating pace of change in business environments;
- knowledge viewed as a source of competitive advantage;
- customers becoming more demanding;
- dissatisfaction with the existing management paradigm; and
- the increasing intensity of competition.

Further, Senge (1999) purported that the real lesson of the quality movement is in the 'learning' that organisations obtain from the process of deployment. In addition, he demonstrated that the evolution of learning organisations can be studied as a series of three quality waves:

- First quality wave: where the primary focus of change was on front-line workers.
- Second quality wave: based on improving management's effectiveness.
- Third quality wave: where learning is institutionalised.

The above points are critical aspects of achieving organisational excellence. The MFAM has been designed to overcome the problematic issues of becoming a 'learning organisation' by the application of 'triple-loop learning'. It has also built in the key components of management, thus overcoming the issues noted above for construction-related companies.

The deployment of the MFAM has to be led (as do all change management processes) by senior management. Therefore, the deployment has to be planned and fully resourced. In fact, it should be subjected to the same monitoring processes as any other change project.

The MFA model recognises that sustainable excellence in all aspects of performance is based on the management functions of forecasting/planning, organising, co-ordinating, motivating, controlling and communication. Excellence is dependent upon balancing and satisfying the needs of all relevant stakeholders (this includes people employed, customers, suppliers and society in general, as well as those with financial interests in the construction organisation).

Construction firms perform more effectively and efficiently when all interrelated activities are understood and systematically managed and decisions concerning current operations and planned improvements are made using reliable information that includes stakeholder perceptions. So the application of RADAR is essential if a truly holistic control mechanism is to be attained.

Corporate excellence is measured by an organisation's ability to both achieve and sustain outstanding results for its stakeholders, thus MFAM linked to RADAR has been developed. The MFAM has found favour with the European Foundation for Quality Management and 'SixSigma', both having published the model.

Questions for the reader

Here follows a number of questions related specifically to the information presented within this chapter. Attempt each question without reference to the chapter in order to assess how much you have learned. The answers are provided at the end of the book.

Question 1

Define the terms 'intrinsic' and 'extrinsic' motivation.

Question 2

State the advantages of adopting a post-modernist philosophy for a construction company.

Question 3

Define the terms *single loop*, *double loop* and *triple loop*, with regard to organisational learning.

Question 4 – Case Study: deploying the MFA model

A managing director of a construction company has decided to fully engage with the concept of a learning organisation. There exists, however, some resistance to this approach within the host company. The managing director thinks that the deployment of the MFAM would be a good starting point for the company. **You** have been appointed as external consultant and asked to prepare a presentation to convince the rest of the board of directors (based upon this chapter). Your presentation is to be titled 'The Organisational Benefits to Be Gained From Deploying the MFAM' and should consist of a list of key bullet points.

When complete, compare your bullet point list with that included at the end of the book.

Further reading

Goh, S.C. (1998). Towards a learning organisation: The strategic building blocks. *SAM: Advanced Management Journal*, 63 (2), pp. 15–20.

Hermel, P.H. (1997). The new focus of total quality in Europe and the US. *Total Quality Management*, 8 (4), pp. 131–143.

McCabe, S. (1998). *Quality Improvement Techniques in Construction*. Edinburgh Gate: Longman.

Oakland, J., Tanner, S., and Gadd, K. (2002). Best practice in business excellence. *Total Quality Management*, 13 (8), pp. 1125–1139.

Patton, R.A., and Mccalman, J. (2000). *Change Management: A Guide to Effective Implementation*. 2nd ed. London: Sage.

Watson, P. (2002). Developing an efficient and effective control system. *Journal of the Association of Building Engineers* (February), 77 (3), pp. 28–29.

References

Argyris, C., and Schön, D.A. (1974). *Theory in Practice: Increasing Professional Effectiveness*. San Francisco: Jossey-Bass.

Argyris, C., and Schön, D.A. (1978). *Organisational Learning: A Theory of Action Perspective*. Reading, MA: Addison-Wesley.

Barlow, J., and Jashapara, A. (1998). Organisational learning and inter-firm 'partnering' in the UK construction industry. [Online]. *The Learning Organisation*, 5 (2), pp. 86–98. Available at: Emerald Fulltext Database [Accessed 9 November 2004].

Beechner, A.B., and Hamilton, K.A. (1999). Infinity model for organizational excellence. *Proceedings of 53rd Annual Quality Congress*. Milwaukee, WI: American Society for Quality, pp. 333–336.

Bessant, J. (1998). Learning and continuous improvement. In Tidd, J. (Ed.), *Measuring Strategic Competencies: Technological, Market and Organizational Indicators of Innovation*. London: Imperial College Press.

Bessant, J, and Caffyn, S. (1997). High involvement innovation. *International Journal of Technology Management*, 14 (1), pp. 7–28.

Bounds, G., Yorks, L., Adams, D.M., and Ranney, G. (1994). *Beyond Total Quality Management: Towards the Emerging Paradigm*. New York: McGraw-Hill International Editions.

Burns, T., and Stalker, G.M. (1961). *The Management of Innovation*. London: Pergammon.

Cole, G. (1995). *Organisational Behaviour. Theory and Practice*. London: Thompson.

Cummings, A. (2005). A critical examination of the key issues that influence a large construction company's ability to become a learning organisation. BSc (Hons) Dissertation, Sheffield Hallam University.

Dahlgaard, S.M.P. (2004). Perspectives on learning—a literature review. *European Quality*, 11 (1), pp. 32–047.

Day, G.S., and Wensley, R. (1988). Assessing advantage: A framework for diagnosing competitive superiority. *Journal of Marketing*, 52, pp. 1–20.

de Geus, A.P. (1988). Planning as Learning. *Harvard Business Review*, 66 (2), pp. 70–74.

European Foundation for Quality Management (1999). *Radar and the EFQM Excellence Model*. EFQM Press Releases & Announcements. [Online]. Available at: www.efgm.org [Accessed 12 June 2000].

Fayol, H. (1949). General and Industrial Management, Sir Isaac Pitman & Sons Ltd., London.

Fryer, B. (1997). *The Practice of Construction Management*. 3rd ed. Oxford: Blackwell Science, Ltd.

Garvin, D.A. (1993). Building a learning organisation. *Harvard Business Review*, 71 (4), pp. 78–91. [Online]. Available at: EBSCO Business Source Premier Database [Accessed 4 November 2004].

Greising, D. (1994). Quality: How to make it pay. *Business Week* (8 August), pp. 54–59.

Griffith, A., and Watson, P. (2004). *Construction Management: Principles and Practice*. Basingstoke: Palgrave Macmillan.

Hamel, G., and Prahalad, C.K. (1994). *Competing for the Future*. Boston: Harvard Business School Press.

Hardy, L. (1983). *Successful Business Strategy: How to Win the Market Place*. London: Kogan Page.

Hermel, P., and Ramis-Pujol, J. (2003). An evolution of excellence: Some main trends. *TQM Magazine*, 15 (4), pp. 230–243.

Ho, S.K.M. (1999). Total learning organisation. *The Learning Organisation*, 6 (3), pp. 116–120. [Online]. Available at: Emerald Fulltext Database [Accessed 4 November 2004].

Hutchin, T. (2001). *Unconstrained Organisations: Managing Sustainable Change: Unlocking the Potential of People Within Organisations*. London: Thomas Telford Ltd.

Imai, K. (1987). *Kaizen*. New York: Random House.

Jackson, N., and Carpenter, P. (2000). *Rethinking Organisational Behaviour.* Harlow: Prentice Hall, Pearson Education Ltd.

Johnson, G., and Scholes, K. (2002). *Exploring Corporate Strategy: Text and Cases.* 6th ed. Harlow: Pearson Education Ltd.

Leonard-Barton, D. (1992). The organisation as learning laboratory. *Sloan Management Review,* 34 (1), pp. 23–38.

Lobermans, J. (2002). Synergising the learning organisation and knowledge management. *Journal of Knowledge Management,* 6, pp. 285–294. [Online]. Available at: Emerald Fulltext Database [Accessed 5 November 2004].

Love, P.E.D. (2004). Nurturing a learning organisation in construction: A focus on strategic shift, organizational transformation, customer orientation and quality centred learning. *Construction Innovation,* 4 (2), pp. 113–126. [Online]. Available at: EBSCO Business Source Premier Database [Accessed 16 November 2004].

Majaro, S. (1992). *Managing Ideas for Profit.* Maidenhead: McGraw-Hill Book Company.

Malhotra, Y. (1996). *Organisational Learning and Learning Organisations: An Overview.* [Online]. Available at: www.kmbook.com/orglrng.htm [Accessed 5 November 2004].

Martin, J., and Cullen, P. (2005). *When Precedents Are Not Enough: Creating Consistent Contracts in a Changing Organisation.* Proceedings of the International Commercial Management Symposium, April 7. The University of Management, UK.

McCabe, S. (2001). *Benchmarking in Construction.* London: Blackwell Science.

Morris, D., and Brandon, J. (1993). *Re-Engineering Your Business.* London: McGraw-Hill.

Naoum, S. (2001). *People and Organisational Management in Construction.* London: Thomas Telford Ltd.

Newell, S. (2004). Enhancing cross-project learning. *Engineering Management Journal,* 16 (1), pp. 12–20. [Online]. Available at: EBSCO Business Source Premier Database [Accessed 16 November 2004].

Ngowi, A.B. (2001). The competition aspect of construction alliances. *Logistics Information Management,* 14 (4), pp. 242–249.

Nonaka, I. (1991). The knowledge: Creating company. *Harvard Business Review,* 69 (6), pp. 96–104. [Online]. Available at: ESBCO Business Source Review Database [Accessed 7 November 2004].

Nyhan, B. (2004). European perspectives on the learning organisation. *Journal of European Industrial Training,* 28 (1), pp. 67–92. [Online]. Available at: Emerald Fulltext Database [Accessed 16 November 2004].

Örtenbland, A. (2004). The learning organisation: Towards an integrated model. *The Learning Organisation,* 11, pp. 129–144. [Online]. Available at: Emerald Fulltext Database [Accessed 4 November 2004].

Pascale, R.T. (1992). *Les risques de l'excellence: La Strategie de conflicts constructifs.* Paris: Inter Editions.

Passmore, W.A. (1994). *Creating Strategic Change.* New York: J. Wiley & Sons Inc.

Pedlar, M., and Aspinwall, K.A. (1998). *Concise Guide to the Learning Organisation.* London: Lemos and Crane.

Pedler, M., Boydell, T., and Burgoyne, J. (1988). *Learning Company Project Report.* Sheffield: Manpower Services Commission.

Peters, T. (1988). Facing up to the need for a management revolution. *California Management Review* (Winter), 30 (2), pp. 7–38.

Revans, R.W. (1983). *ABC of Action Learning.* Bromley: Chartwell Bratt.

Robinson, A. (1991). *Continuous Improvement in Operations.* Cambridge, MA: Productivity Press.

Sayles, L.R. (1989). *Leadership Managing in Real Organisations.* London: McGraw-Hill, London.

Schroeder, M., and Robinson, A. (1993). Training, continuous improvement and human relations: The US TWI programs and Japanese management style. *California Management Review*, 35 (2), pp. 35–57.

Senge, P.M. (1990). *The Fifth Discipline: The Art and Practice of the Learning Organisation*. London: Century Business.

Senge, P.M. (1999). Association for Quality & Participation. *It's the Learning: The Real Lesson of Quality Movement* (November–December), 22 (6), pp. 34–40.

Shakantu, W., and Talukhaba, A. (2002). *Benchmarking Best Practice to Achieve a Competitive Advantage in the South African Construction Industry*. In Ahmed, S.M., Ahmad, I., Tang, S.I., and Azhar, S. (Eds.). First International Conference on Construction in the 21st Century "Challenges and Opportunities in Management and Technology" (CITC-I), April 25–26, 2002, Florida International University, Miami, USA.

Shukla, M. (1997). *Competing through Knowledge: Building a Learning Organisation*. New Delhi: Response Books.

Stata, R. (1989). Organisational learning—The key to management innovation. *MIT Sloan Management Review*, 30 (3), p. 63.

Sun, P.Y.T., and Scott, J.L. (2003). Exploring the divide: Organisational learning and learning organisation. *The Learning Organisation*, 10 (4), pp. 2002–2215. [Online]. Available at: Emerald Fulltext Database [Accessed 4 November 2004].

Teece, D., and Pisano, G. (1994). The dynamic capabilities of firms: An introduction. *Industrial and Corporate Change*, 3 (3), pp. 537–555.

Tidd, J., Bessant, J., and Pavitt, K. (1997). *Managing Innovation: Integrating Technological, Organizational and Market Change*. Chichester: John Wiley.

Watson, P., and Chileshe, N. (1998). Aspects of Total Quality Management (TQM) implementation within a construction operational environment. *South African First Congress on Total Quality Management in Construction* (November), pp. 94–101 .

Watson, P., Chileshe, N. and Maslow, D. (2005a). *Addressing Sustainable Competitive Advantage via a Functional Assessment Model*. International Commercial Management Symposium. 7 April. University of Manchester, UK.

Watson, P., Maslow, D., and Chilshe, N. (2005). Management assessment for competitive advantage. *iSixSigma Insights* (March 7), 6 (19).

Weber, M., 1908 (1968). *Economy and Society*. Translated and Edited by Roth, G., and Witrich, C. New York: Irving Publications.

6

QUALITY MANAGEMENT SYSTEMS FOR HEALTH AND SAFETY IN CONSTRUCTION

Introduction

Excellence in health and safety management is an essential attribute of successful modern-day construction organisations. Poor health and safety management can impact upon a construction organisation's reputation, the timely progress of its projects, the morale and commitment of its workforce and the size and future surety of its order book.

This chapter serves to inform of occupational health and safety management systems and outlines the essential components of such systems for organisations. Advocated benefits and problems associated with occupational health and safety management systems are indicated and differing standards and guidance documents are introduced. Examples of useful documentation for contributing to the systematic management and audit of health and safety on construction projects are provided at the end of the chapter.

Learning outcomes

Upon completion of this chapter, the reader will be able to demonstrate an understanding of:

- Essential components of occupational health and safety management systems (OHSMS).
- Advocated benefits and problems associated with occupational health and safety management systems.
- Different health and safety management standards and guidance documents.
- Issues associated with developing an OHSMS within an organisation.
- The role of inspection and audit as tools and processes that help reduce health and safety risks within the construction workplace.

Essential components of occupational health and safety management systems

Like many management models, OHSMS are commonly founded upon Deming's dynamic control loop cycle. This cycle is illustrated in *Figure 6.1*.

The Institution of Occupational Safety and Health (IOSH) (2009) expresses the key components of occupational health and safety management systems in terms of a Deming 'Plan-Do-Check-Act' diagram, as illustrated in *Figure 6.2*.

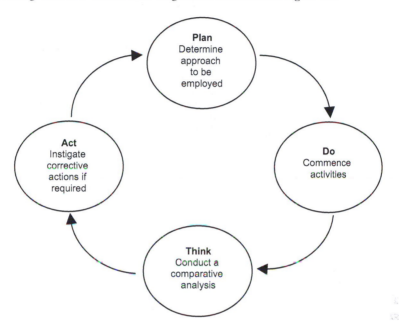

FIGURE 6.1 Deming's dynamic control loop cycle

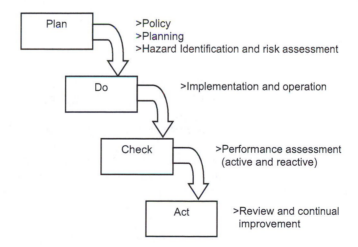

FIGURE 6.2 Key components of the OHSMS, according to IOSH

IOSH (2009) also suggests that effective health and safety management systems contain the following elements:

- Policy – The organisation's statement of commitment and vision. Senior management must lead this policy and be accountable for it.
- Planning – This should address how legal requirements are identified, how hazards are identified and the resultant risks assessed and controlled. It should also document preparation regarding planning for and responding to emergencies.
- Organising – The organisation's structure needs to be defined and health and safety clearly allocated in a manner linked to operational controls. Furthermore there needs to be ways of delivering and ensuring awareness, competence, consultation and training.
- Workers/employee representatives – Such representation can invaluably facilitate the health and safety management of the organisation, particularly with regard to improvement opportunities and risk management.
- Communicating – It is essential that this is two way between the organisation's managers and workers, is regular and ongoing and includes health and safety information relating to work procedures and all aspects of the organisation's OHSMS.
- Consulting – All stakeholders of the organisation need to be identified and consulted effectively regarding health and safety in order to proactively access their knowledge, views, requirements and expertise as well as the reactive feedback.
- Implementing and operating – The OHSMS needs to be put into practice, in its entirety.
- Measuring performance – This can be undertaken by evaluating data relating to incidents (accidents and near misses) and ill health as well as information obtained from, amongst other things, hazard identification, risk assessments, regular inspections, health and safety committees and training activities.
- Corrective and preventive actions – There must be a systematic approach to proactively prevent incidents and accidents ill health, as well as corrective measures implemented from the investigation of incidents, accidents and ill health.
- Management review – This must be done in order to ensure compliance with legal requirements and appraise the performance achieved against objectives set and re-evaluate the system itself and its resourcing.
- Continual improvement – A commitment to proactively manage health and safety risks is at the core of the system in order to effectively reduce incidents of ill health and accidents with the efficient deployment of reduced resources.

The United Kingdom's Health and Safety Executive (HSE) documents the essential components of a successful health and safety management system in the publication *Successful Health and Safety Management*. The HSE outlines the following as being essential components of successful health and safety management:

- A clear policy for health and safety.
- Organisation of all employees for the management of health and safety.

- Planning for health and safety – via setting objectives and targets, identifying hazards, assessing risks and establishing standards for the performance of the organisation to be measured against.
- Measurement of health and safety performance.
- Informed improvement by auditing and reviewing safety performance and practice.

Figure 6.3 illustrates the essential components of successful health and safety management as identified by the HSE.

The International Labour Office (ILO) has also identified key components of an OHSMS. The ILO considers that an organisation's system should contain policy, organising, planning and implementation, evaluation and action for improvement. These components are required to be structured in a manner rooted in Deming's dynamic control loop cycle, as illustrated in *Figure 6.4*.

The European Agency for Safety and Health at Work (2002) suggests that an ideal occupational health and safety management system should include a number of key processes. *Table 6.1* presents these key processes.

A further occupational health and safety management system is provided by the Occupational Health and Safety Assessment Series (OHSAS) 18001 management systems standard. This standard is commonly recognised around the globe as a leading health and safety management systems standard. Section 4.1 of OHSAS

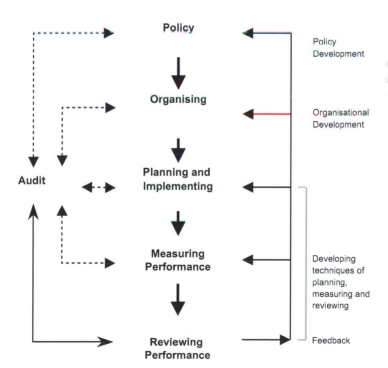

FIGURE 6.3 Key components of successful health and safety identified by the HSE

Develop Policy → Organise

Regular, systematic audit for continuous improvement

Improvement actions

Implement

Review and evaluate

FIGURE 6.4 Key components of the OHSMS, according to the International Labour Office

TABLE 6.1 Key processes of an occupational health and safety management system, according to the European Agency for Safety and Health at Work (2002)

OHS Input – Initiation
- Management commitment and resources
- Regulatory compliance and system conformance
- Accountability, responsibility and authority
- Employee participation

OHS Process

Implementation	Formulation
• Training system • Hazard control system • Preventive and corrective action system • Procurement and contracting	• OHS Policy / goals and objectives • Performance measures • System planning and development • Baseline evaluation and hazard assessment • OHSMS manual and procedures

---

OHS Output Feedback
- OHS goals and objectives
- Illness and injury rates
- Workforce health
- Changes in efficiency
- Overall performance of the organisation

OHS Feedback Evaluation
- Communication system – including document and record management system
- Evaluation system – auditing and self inspection, incident investigation, root-cause analysis, health/medical programme and surveillance

---

Open Systems Elements
- Continuous improvement
- Management review
- Integration

18001:2007 sets out five general requirements for an organisation's occupational health and safety management system, these being:

1 Establish a management system
2 Document the management system
3 Implement the management system
4 Maintain the management system
5 Continually improve the management system.

Benefits and problems associated with OHS management systems

Quantifying the benefits and problems associated with the establishment and delivery of an organisation's OHSMS can be challenging. A key qualitative benefit that can be readily associated with the implementation of an OHSMS, though, is the very visible, strategic and operational, *commitment* provided to the health, safety and well-being of the organisation's workforce community. An OHSMS can also assist in the delivery of a number of the organisation's legal and moral obligations and can also facilitate an internal focus on good management practice and continuous improvement.

A number of benefits and problematic issues can be associated with the deployment of OHS management systems. Benefits can include:

- Improved prevention of occupational injury and disease – a safer and healthier workplace.
- The provision of a framework for identifying hazards and managing the resultant risks.
- A reduction in the loss of working days due to accidents and injury.
- A reduction in the incidences of employee compensation claims.
- The development of a reviewable approach for meeting legislative requirements, duties of care and due diligence.
- A reduction in insurance premiums.
- Improved morale and productivity brought about by employee inclusivity with developing and running.
- Enhanced working methods that facilitate improvement in production and productivity rates.
- Enhanced reputation of the organisation with a visible and tangible commitment to continuous improvement and inclusive, consultative management mechanism.
- Reduced staff turnover and thereby reduced 'replacement costs'.
- Improved ability to attract skilled personnel.
- Improved commercial potential – inclusion on tender lists is increased as the potential for meeting the pre-qualification requirements of significant clients is enhanced.

These benefits can be paralleled with those associated with the implementation of a workplace health promotion programme within an organisation. *Figure 6.5*

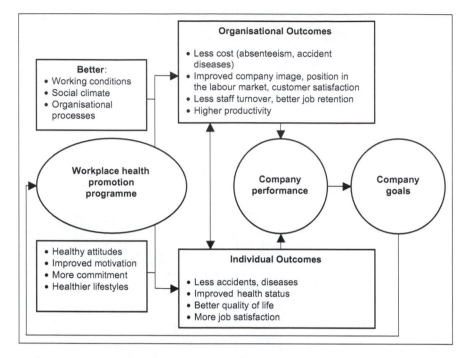

FIGURE 6.5   Framework illustrating the effects and outcomes of workplace health promotion

illustrates the effects and outcomes of workplace health promotion and is derived from the European Agency for Safety and Health at Work (2009).

Problems that can be associated with the development and delivery of an OHSMS include:

- The system is not organisation specific. The system needs to be tailored to the organisation and its culture. There is no ready-made 'off the shelf' solution.
- Management support may be lacking. The leadership and commitment of management needs to be visible and suitable management priority needs to be given to establishing, developing and improving the system.
- Understanding of the purpose and benefits of the system is lacking – here the system can be viewed as a 'paper trail' or hindrance to daily work, potentially with procedures put in place in a top-down manner.
- The OHSMS is established due purely to external drivers – possibly in order to enable inclusion on client tender lists. With this external driver alone, the system will rarely achieve the necessary 'ownership' by those internal to the organisation.
- There is insufficient 'ownership' of the system from persons across the organisation. Without broad and effective participation in development, sustained delivery and improvement, the system will be viewed as one that is 'imposed'.

Health and safety management standards and guidance documents

Common to the various occupational health and safety management standards and guidance documents that have been developed is that each is based upon Deming's dynamic control loop cycle. Furthermore each shares the intent of facilitating the delivery of robust, systemised occupational health and safety management practice.

When undertaking to develop and implement an occupational health and safety management system within an organisation, a number of 'standards' are worthy of consideration and consultation; these include:

- HSG65 Successful Health and Safety Management
- ILO OSH:2001 Guidelines on occupational health and safety management systems
- OHSAS 18001:2007 Occupational health and safety management systems – requirements
- OHSAS 18002:2008 Occupational health and safety management systems – guidelines for the implementation of OHSAS 18001:2007
- BS 18004:2008 Guide to achieving effective occupational health and safety performance.

Knowledge of each of these occupational health and safety standards is worthwhile, especially when undertaking to develop and implement an occupational health and safety management system. The standards do not prescribe what an organisation must do, instead the standards provide a 'framework' to help key issues to be identified in order that the system that is developed is suitably and effectively aligned with the host organisation.

Figure 6.6 presents a timeline overview of the development of various key occupational health and safety management standards.

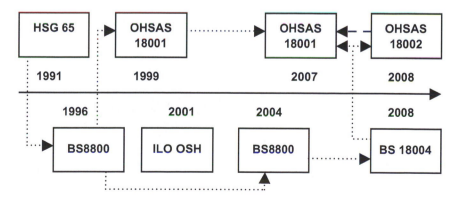

FIGURE 6.6   Timeline of occupational health and safety management standards

HSG65 Successful Health and Safety Management

Successful Health and Safety Management was initially prepared by the Health and Safety Executive's Accident Prevention Advisory Unit to provide guidance for directors, managers and safety professionals who were seeking to improve health and safety performance. It was first published in 1991 and has since been revised. It is not an approved code of practice or a certifiable international standard. It is a guidance document.

In presenting sound guidance on the practice of health and safety management within organisations, the following content is addressed:

- Effective health and safety policies
- Organising for health and safety
- Planning and implementing
- Measuring performance
- Auditing and reviewing performance.

The POPMAR model for managing health and safety is a well-recognised feature of this systematic guidance. Here policy, organisation, planning and implementing, measuring, auditing and reviewing are presented as the key elements of a successful health and safety management system.

ILO OSH:2001 Guidelines on Occupational Health and Safety Management Systems

The ILO issued ILO OHS Guidelines on Occupational Health and Safety Management System in 2001, further toward consultation and a meeting of experts in April 2001. The guidelines are intended for application at two levels, a national level and an organisational level. The organisational level seeks to:

a   Provide guidance regarding the integration of OHS management system elements in the organisation as a component of policy and management arrangements; and
b   Motivate all members of the organisation, particularly employers, owners, managerial staff, workers and their representatives, in applying appropriate OHS management principles and methods to continually improve OHS performance (ILO OHS 2001).

The guidelines are not legally binding and their application does not necessitate certification. The following content is addressed by the guidelines:

- Policy
- Organising
- Planning and implementation
- Evaluation
- Action for Improvement.

OHSAS 18001:2007 Occupational health and safety management systems – requirements

OHSAS 18001 documents requirements for an occupational health and safety management system and serves to support organisations control OH&S risks and improve performance. OHSAS 18001:2007 replaces OHSAS 18001:1999.

The standard has been adopted by British Standards. BS OHSAS 18001 states the requirements that must be met to demonstrate that an organisation has an effective occupational health and safety management system. It is a standard for which certification for compliance can be sought. This enables an organisation to demonstrate via independent audit to their stakeholders that the organisation operates a health and safety management system with elements and characteristics that are in accordance with the standard.

The standard is supported by OHSAS BS 18002:2008 Guidelines for the Implementation of OHSAS 18001:2007 and BS 18004:2008 Guide to Achieving Effective Occupational Health and Safety Performance.

Baker (2001) argues that whilst organisations can see the value of an OHSMS such as OHSAS 18001, there can be perceived limitations associated with such a standard. The certification of OHSAS 18001, it can be argued, could indicate a good level of safety administration rather than effective safety and health management – it can be regarded as focusing on written documentation.

Table 6.2 provides an overview of the content of this safety standard.

TABLE 6.2 Overview of OHSAS 18001:2007

1	Scope
2	Reference publications
3	Terms and definitions
3.1	Acceptable risk
3.2	Audit
3.3	Continual improvement
3.4	Corrective action
3.5	Document
3.6	Hazard
3.7	Hazard identification
3.8	Ill health
3.9	Incident
3.10	Interested party
3.11	Non-conformity
3.12	Occupational health and safety (OH&S)
3.13	OH&S management system
3.14	OH&S objective
3.15	OH&S performance
3.16	OH&S policy
3.17	Organisation

(Continued)

TABLE 6.2 (Continued)

3.18	Preventive action
3.21	Risk
3.22	Risk assessment
3.23	Workplace
4.1	General requirements
4.2	OH&S policy
4.3.1	Hazard identification, risk assessment and determining controls
4.3.2	Legal and other requirements
4.3.3	Objectives and program(s)
4.4.1	Resources, roles, responsibility, accountability and authority
4.4.2	Competence, training and awareness
4.4.3.1	Communication
4.4.3.2	Participation and consultation
4.4.4	Documentation
4.4.5	Control of documents
4.4.6	Operational control
4.4.7	Emergency preparedness and response
4.5	Checking
4.5.1	Performance measurement and monitoring
4.5.2	Evaluation of compliance
4.5.3	Incident investigation, nonconformity, corrective action and preventive action
4.5.3.1	Incident investigation
4.5.3.2	Nonconformity, corrective and preventive action
4.5.4	Control of records
4.5.5	Internal audit
4.6	Management review

OHSAS 18002:2008 Occupational health and safety management systems – guidelines for the implementation of OHSAS 18001:2007

These guidelines serve to support the implementation of OHSAS 18001 and explain the principles of OHSAS 18001. This standard serves to assist organisations to understand and implement OHSAS 18001 through the provision of examples and aspects to consider when undertaking to implement or audit OHSAS 18001.

BS 18004:2008 Guide to achieving effective occupational health and safety performance

This standard replaces BS8800 2004 and serves to add to the requirements of 18001 and the guidance in 18002 by providing more detailed information about key elements of effective occupational health and safety management. BS 18004 is for organisations that seek to:

- Establish an OH&S management system to control risks to personnel and other interested parties who could be exposed to OH&S hazards associated with its activities.
- Implement, maintain and continually improve the OH&S management system.
- Demonstrate commitment to good practice, including self-regulation and continuous improvement in OH&S performance.
- Assure conformity with the H&S policy and BS OHSAS 18801 by either self-determination and a declaration, seeking confirmation from either an organisational stakeholder or an external party or via certification of the OHSMS by an external organisation.

Furthermore the standard also provides guidance on promoting an effective OH&S management system and investigating hazardous events.

Developing an occupational health and safety management system

When undertaking to develop and implement an OHSMS such as OHSAS 18001:2007, it is necessary for the host organisation to rigorously align the H&S standard with the organisation. This tailored alignment necessitates considered documentation, so as to provide evidence of the critical process of the constructive application of the standard to the host organisation. This process requires the identification of H&S aspects or 'inputs to the system'. The Institution of Occupational Safety and Health (IOSH) outlines a process for the development within an organisation of an occupational health and safety management system. This process is arranged into three sections, these being 'typical inputs', a 'gap analysis review' and the 'development of a draft management system' that documents a number of key components. IOSH's process for developing an OHSMS is presented in *Figure 6.7*.

Developing documentation – a case study example

When undertaking to develop and document a management system, a 'consensus chart' is proposed by Laman (2009) in order to classify issues and improve the process. This chart is presented in *Figure 6.8* and is completed with safety and health issues of one example organisation completed within the chart matrix. Application of this consensus chart approach is considered to facilitate a standardised and effective process for obtaining buy-in and comprehensive documentation. Laman considers the chart to be especially useful in an environment such as that of an organisation's OHSMS with the following characteristics:

- Continuous improvement
- Differences of opinion
- Desire to optimise documentation

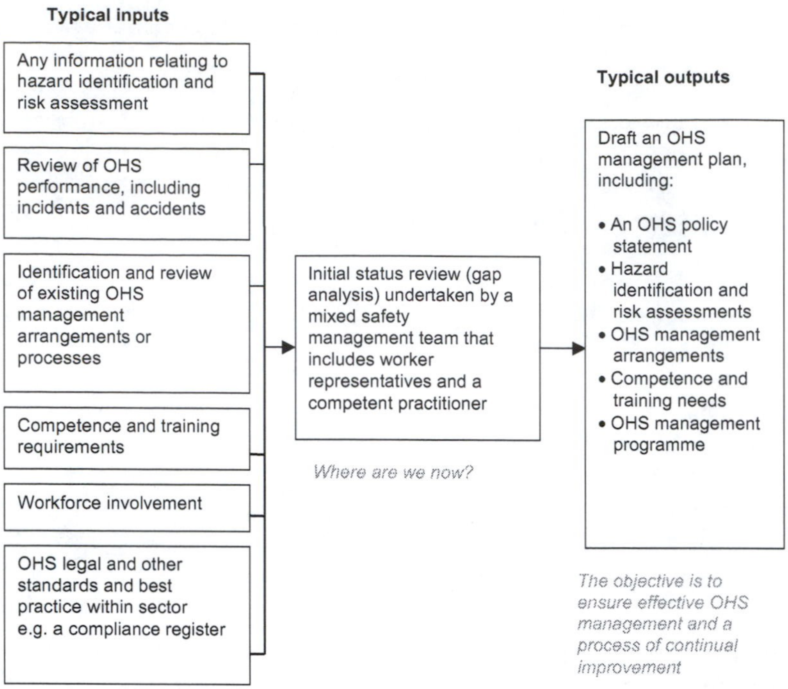

Figure 6.7 content

Typical inputs

- Any information relating to hazard identification and risk assessment
- Review of OHS performance, including incidents and accidents
- Identification and review of existing OHS management arrangements or processes
- Competence and training requirements
- Workforce involvement
- OHS legal and other standards and best practice within sector e.g. a compliance register

Initial status review (gap analysis) undertaken by a mixed safety management team that includes worker representatives and a competent practitioner

Where are we now?

Typical outputs

Draft an OHS management plan, including:

- An OHS policy statement
- Hazard identification and risk assessments
- OHS management arrangements
- Competence and training needs
- OHS management programme

The objective is to ensure effective OHS management and a process of continual improvement

FIGURE 6.7 Process for developing an OHSMS

(IOSH 2009 pp 23)

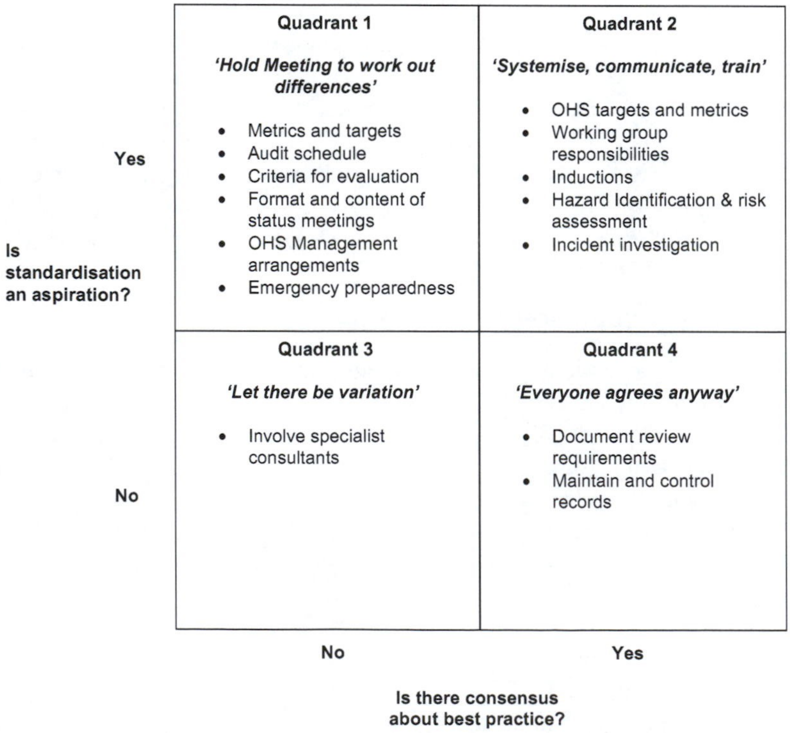

Figure 6.8 content

Is standardisation an aspiration?

Yes

Quadrant 1

'Hold Meeting to work out differences'

- Metrics and targets
- Audit schedule
- Criteria for evaluation
- Format and content of status meetings
- OHS Management arrangements
- Emergency preparedness

Quadrant 2

'Systemise, communicate, train'

- OHS targets and metrics
- Working group responsibilities
- Inductions
- Hazard Identification & risk assessment
- Incident investigation

No

Quadrant 3

'Let there be variation'

- Involve specialist consultants

Quadrant 4

'Everyone agrees anyway'

- Document review requirements
- Maintain and control records

No   **Yes**

Is there consensus about best practice?

FIGURE 6.8 Consensus chart (adapted from Laman)

- Appreciation of the team approach
- Complex processes that interact with other processes

In applying the 'consensus chart' approach to the documentation development process, a representation of management and employees is necessary. This representation is required to give due consideration to identifying health and safety issues and the current context of each issue within the organisation.

Laman suggests posing two questions during this process:

1 Do we want to standardise?
2 Is there consensus about best practices?

The answers, or 'issue responses', provided by the representative group are located within one of the four 'action' quadrants of the consensus chart.

A **'Hold meetings to work out differences'** – Here standardisation is desired, but consensus has not yet been achieved.
B **'Systemise, communicate, train'** – There is both desire and consensus to standardise. Process improvements can be initiated.
C **'Let there be variation'** – There is no desire to standardise and consensus is not held. No action results from this outcome.
D **'Everyone agrees anyway'** – Consensus exists but there is no need to standardise.

Where standardisation is agreed – 'Systemise, communicate, train' is the outcome – new documentation will result. This documentation could include new and revised policies, procedures, forms and training plans.

Measuring performance of an organisation's occupational health and safety management system

The development and implementation of an OHSMS can bring about reductions in incidents, accidents and ill health. Furthermore it can reduce resultant costs, lost time and insurance premiums as well as improve reputation, well-being and motivation. This is all well and good but can lead to organisations focusing attention upon metrics that narrowly concern safety performance resulting or trailing from the OHSMS. It is imperative that the performance of the system itself is measured and monitored in terms of metrics that seek to deliver efficiency and continuous improvement.

Warren (2005 in European Agency for Safety and Health at Work 2009) suggests that the performance measurement of management systems must be 'SMART': specific, measurable, achievable, relevant and time-based.

Specific:	Performance criteria should be as specific as possible to make sure that it is easy to identify what is being measured.
Measurable:	Performance criteria need to be measurable, either in quantity or by quality, to check that stipulated goals are being met.

Achievable: Unrealistic goals may cause disease within an organisation. However, the challenge of goals that stretch an organisation a little may be beneficial.

Relevant: The performance measurements should be relevant to the organisation's overall mission and to the strategic objectives of any programme.

Time-based: The performance measurements should be achievable within a specific period.

These five performance measurement characteristics provide helpful direction for the development of performance metrics for a safety management system.

Street (2000) suggests that performance metrics can be categorised into three types, these being 'trailing indicators', 'current indicators' and 'leading indicators', as illustrated in *Table 6.3*. Together these three categories of metrics provide a comprehensive platform of data and information from which a safety management system can be viewed and evaluated.

When measuring the performance of an OHSMS it is important to appreciate the distinction between 'incidents' and 'nonconformities'. Section 4.5.3 of OHSAS

TABLE 6.3 Three types of organisational H&S performance measurement indicators

Trailing Indicator	*Current Indicator*	*Leading Indicator*
• Injury and illness statistics	• Safe and unsafe acts indices	• Audit programme quality, including adherence to scheduling
• Litigation costs	• Incident investigation reporting and analysis	• Volume of repeat injuries
• Disability costs	• Serious potential incident frequency	• Analysis of process hazards reviews
• Workers' compensation costs	• Safety audit findings	• Number of safety work orders/unit of time
• Vehicle accident statistics	• Occupational medical visits	• The process of Incident reporting, investigation and follow up
• Regulatory citation and penalties	• Volume and accuracy of training records and resultant effectiveness of training	• Attitudes and perceptions of employees
• Process release statistics	• Action on past employee surveys	• Employee safety suggestions – quality and quantity
	• Safety meetings – attendance at and issues addressed	• Involvement of senior management and employees in safety processes and systems

(adapted from Street 2000)

18001:2007 concerns "incident investigation, nonconformity, corrective action and preventative action". OHSAS 18001 defines an incident as a "work-related event in which an injury or ill health or fatality occurred, or could have occurred". An 'accident' is a particular category of incident, one in which injury or illness results, regardless of severity. Another category of incident is one where no illness or injury results; this is termed a 'near miss'.

It is possible for an organisation to be operating in a manner of nonconformity with its OHSMS. Here aspects such as work procedures, information flows, worker engagement and consultation, training, risk assessment or the like may not be in full accordance with the developed OHSMS. An organisation may have nonconformities without the immediate occurrence of incidents. Such nonconformities indicate insufficient attention to implementation, control, review and ownership of the system.

Auditing the system

Audit is a necessary component of any system that seeks to ensure conformity and continuous improvement. The performance metrics utilised for the evaluation of the system can provide a worthwhile data set for the monitoring and managed improvement of the organisation's safety management system. The Health and Safety Executive prescribe a worthwhile checklist tool for the auditing of eight components of an organisation's health and safety management system (HSE 2008). The tool enables an organisation to assess its health and safety management system by 'self-scoring' against specified elements. By repeating the self-assessment exercise after improvements have been made to the system, progress can be measurably scored and recorded over time. *Table 6.4* presents this self-assessment audit tool.

TABLE 6.4 Self-assessment audit checklist

AUDIT OF SAFETY POLICY

	Fully Met (Score 2)	Partially Met (Score 1)	Not Met at All (Score 0)
1 The company has a clear, written policy for health and safety at work, signed, dated and communicated to all employees?			
2 The Directors regard health and safety of employees as an important business objective?			
3 The Directors are committed to continuous improvement in health and safety (reducing the number of injuries, cases of work-related ill health, absences from work and accidental loss)?			
4 A named Director or Senior Manager has been given overall responsibility for implementing our health and safety policy?			

(Continued)

TABLE 6.4 (Continued)

	Fully Met (Score 2)	Partially Met (Score 1)	Not Met at All (Score 0)
5 Our policy commits the Directors to preparing regular health and safety improvement plans and regularly reviewing the operation of our health and safety policy?			
6 Our policy encourages the involvement of employees and safety representatives in the health and safety effort?			
7 Our policy includes a commitment to ensuring that all employees are competent to do their jobs safely and without risks to health?			

AUDIT OF ORGANISATION FOR SAFETY CONTROL

1 In our company, responsibilities for all aspects of health and safety have been defined and allocated to our managers, supervisors and team leaders?			
2 Our managers, supervisors and team leaders accept their responsibilities for health and safety and have the time and resources to fulfil them?			
3 Our managers, supervisors and team leaders know what they have to do to fulfil their responsibilities and how they will be held accountable?			
4 We have identified the people responsible for particular health and safety jobs including those requiring special expertise (e.g. our health and safety advisor)?			

AUDIT OF ORGANISATION OF SAFETY COMPETENCE

1 We have assessed the experience, knowledge and skills needed to carry out all tasks safely?			
2 We have a system for ensuring that all our employees, including managers, supervisors and temporary staff, are adequately instructed and trained?			
3 We have a system for ensuring that people doing particularly hazardous work have the necessary training, experience and other qualities to carry out the work safely?			
4 We have arrangements for gaining access to specialist advice and help when we need it?			

	Fully Met (Score 2)	Partially Met (Score 1)	Not Met at All (Score 0)

5 We have systems for ensuring that competence needs are identified and met whenever we take on new employees, promote or transfer people or when people take on new health and safety responsibilities, e.g. when we restructure or reorganise?

AUDIT OF ORGANISATION OF WORKFORCE SAFETY INVOLVEMENT

1 We consult our employees and employee safety representative on all issues that affect health and safety at work?

2 We have an active health and safety committee that is chaired by the appropriate Director or Senior Manager and on which employees from all departments are represented?

3 We involve the workforce in preparing health and safety improvement plans, reviewing our health and safety performance, undertaking risk assessments, preparing safety-related rules and procedures, investigating incidents and problem solving?

4 We have arrangements for cooperating and co-ordinating with contractors and employment agencies whose employees work on our site on health and safety matters?

AUDIT OF ORGANISATION OF SAFETY COMMUNICATION

1 We discuss health and safety regularly and health and safety is on the agenda of management meetings and briefings?

2 We provide clear information about the hazards and risks and about the risk control measures and safe systems of work to people working on our site (which is easily accessible in the relevant work area)

3 Our directors, managers and supervisors are open and approachable on health and safety issues and encourage their staff to discuss health and safety matters?

4 Our Directors, Managers and Team Leaders communicate their commitment to health and safety through their behaviour and by always setting a good example?

(Continued)

TABLE 6.4 (Continued)

AUDIT OF SAFETY PLANNING AND IMPLEMENTATION

	Fully Met *(Score 2)*	*Partially Met* *(Score 1)*	*Not Met at All* *(Score 0)*
1 We have a system for identifying hazards, assessing risks and deciding how they can be eliminated or controlled?			
2 We have a system for planning and scheduling health and safety improvement measures and for prioritising their implementation depending on the nature and level of risk?			
3 We have arrangements for agreeing on measurable health and safety improvement targets with our managers and supervisors?			
4 Our arrangements for purchasing premises, plant, equipment and raw materials and for supplying our products take health and safety into account at the appropriate stage, before implementation of the plan or activity?			
5 We take proper account of health and safety issues when we design processes, equipment, procedures, systems of work and tasks?			
6 We have health and safety rules and procedures covering the significant risks that arise in our day-to-day work activities, including normal production, foreseeable abnormal situations and maintenance work?			
7 We have procedures for dealing with serious and imminent dangers and emergencies?			
8 We set standards against which we can measure our health and safety performance?			

AUDIT OF SAFETY PERFORMANCE MEASUREMENT

1 We have arrangements for monitoring progress with the implementation of our health and safety improvement plans and for measuring the extent to which the targets and objectives set under those plans have been achieved?			
2 We have arrangements for active monitoring that involve checking to ensure that our control measures are working properly, our health and safety rules and procedures are being followed and the health and safety standards we have set for ourselves are being met?			

	Fully Met *(Score 2)*	Partially Met *(Score 1)*	Not Met at All *(Score 0)*

3 We have arrangements for reporting and investigating accidents, incidents, near misses and hazardous situations?

4 Where the arrangements in 2 and 3 above show that controls have not worked properly, our health and safety rules or procedures have not been followed correctly or our safety standards have not been met, we have systems for identifying *why* performance was substandard?

5 We have arrangements for dealing effectively with situations that have created risk, with priority being given where the risks are greatest?

6 We have arrangements for analysing the causes of any potentially serious events so as to identify the underlying root causes, including causes arising from shortcomings in our safety management system and safety culture?

AUDITING OF REVIEWING SAFETY

1 We have regular audits of our safety management system carried out by competent external auditors or competent auditors employed by our company who are independent of the department they are auditing?

2 We use the information from performance monitoring and audits to review the operation of our safety management system and our safety performance?

3 We regularly review how well we have met the objectives in our health and safety improvement plans and whether we have met them in the agreed upon timescales?

4 We analyse the information from performance measurement and use it to identify future improvement targets and to identify particular causes of accident, ill health or poor control of risk to target for future risk reduction effort?

5 We benchmark the performance of our safety management system against that of other businesses in the same industrial sector and/or to monitor our own overall improvement over time?

The H&S management system and the construction project

Building and infrastructure project workplaces are dynamic and diverse in many respects. They are procured in a variety of ways, have various contractual arrangements and differing project management organisational structures and present many technical innovations and challenges. It is imperative that a project health and safety management system can fully address the challenging various arrangements and aspects of differing projects in a systematic manner. The principles of approach for a project safety management system are the very same as for the development and implementation of an organisation-wide occupational health and safety management system: a 'plan, do, check, act' approach is required. Indeed the development and implementation of a project-based approach to health and safety management is a constituent part of a construction organisation's health and safety management system.

Within the UK, the development and implementation of construction project safety management systems is greatly informed by the Construction Design and Management (CDM) Regulations 2015. These regulations require a planned and considered approach throughout the design, delivery, maintenance and decommissioning phases of construction projects. One specific CDM requirement is for the preparation and use of a *construction phase plan*. This is a key component of the construction project safety management system in the UK and requires a construction project to plan and co-ordinate safety management. The construction phase plan is a key component of a principal contractor's construction project safety management system, as indicated in *Figure 6.9*.

Examples of construction project safety management documentation

A useful tool for helping to make construction projects safer has been developed by the Specialist Engineering Contractor's (SEC) Group. The SEC Group have developed a 'Safe Site Access Certificate' which assists in:

- establishing a clear line of communication and mutually agreed criteria for site safety before the work starts;
- helping to make the work safer by reducing, or removing altogether, the risks arising from poor conditions on site; and
- providing a consistent approach to site safety through helping all parties to meet their health and safety responsibilities.

The SEC Group recommend that the checklist certificate be completed jointly contractors and the principal contractor. A copy of the Safe Site Access Certificate is provided in *Table 6.5*

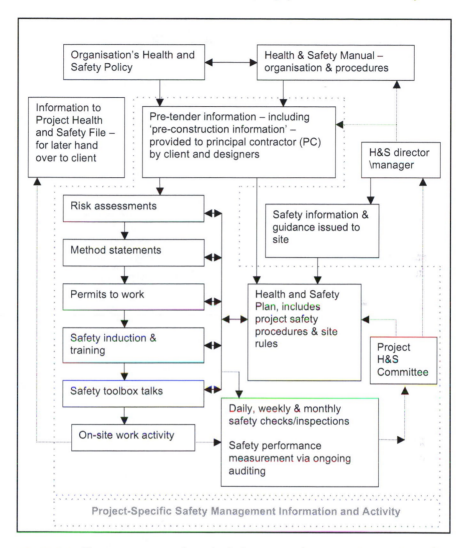

FIGURE 6.9 Key components of a principal contractor's construction project safety management system.

Example of a construction project inspection report form

The implementation of the 'do, check, act' elements of a construction project's safety management system requires the principal contractor to continuously review and audit health and safety related documentation, communications and the physical site environment. It is good practice to document these audit inspections. *Table 6.6* presents a 'project health and safety inspection report form' for use when documenting a review of a construction project's safety documentation, communication and the site environment.

TABLE 6.5 Safe Site Access Certificate

The SEC Group (Revision November 2008)

Contract Information

Contract Name:	Brief details of the contractor's work:
Principal Contractor:	
Contractor:	
Site Address:	Areas of the site the contractor will work in:

Site confirmed as safe and suitable for work by the principal contractor	Site accepted as safe and suitable for work by the contractor *This must be re-checked at the time of starting work on the site – see page 4.*
Name: Position:	Name: Position:
Signature: Date: Time:	Signature: Date: Time:

Provision of Information

	No. Weeks	No. Days	Details / Comments (in general terms)
1. Minimum amount of time before start of construction / installation for planning and preparation [REG.22 (1) (f)]	☐	☐ ▶	

	NO	YES	
2. Has the Principal Contractor issued the part(s) of the **_construction phase plan_** relevant to the work to be carried out? [REG.22 (1) (g)]	☐	☐ ▶	

General hazards [REG.22 (1) (i)]

	N/A	NO	YES	Details / Comments (in general terms)
3. Has the Principal Contractor reported the known significant hazards to the Contractor? (e.g. presence of asbestos containing materials, etc.)	☐	☐	☐ ▶	
4. Has the Principal Contractor given details and locations of all fixed site hazards to Contractor? (e.g. deep water, microwave dishes, contaminated ground, etc.)	☐	☐	☐ ▶	
5. Have other suspected or possible significant hazards been advised to the Contractor? (e.g. work by other contractors, such as lifting operations)	☐	☐	☐ ▶	
6. Are there any other site-specific hazards?	☐	☐	☐ ▶	

Site access & storage

	N/A	NO	YES	Details / Comments (in general terms)
7. Is there clear, adequate and safe access to areas where the Contractor has to work? [REGS 26 & 27] (i.e. free from slipping, tripping and falling hazards, etc.)	☐	☐	☐ ▶	
8. Has the Principal Contractor supplied suitable and sufficient site access lighting and power supplies? [REG 44]	☐	☐	☐ ▶	
9. Are the emergency escape routes clear, suitably marked and provided with emergency lighting where necessary? (i.e. a minimum of 5 lux of lighting from battery operated units) [REG 40 (3)]	☐	☐	☐ ▶	

10. Have overhead and underground services and/or obstructions on the site been identified and marked?
(e.g. cables, manholes, voids, etc affecting access routes, etc) [REG 34]

☐   ☐   ☐ ▶

11. Is hard standing and space available for the delivery and off-loading of huttage, materials, etc, together with easily reached secure storage for materials and/or equipment? [REG.22 (1)]

☐   ☐   ☐ ▶

Contractor's personnel	N/A	NO	YES	Details / Comments (in general terms)

12. Has a site induction, on site-specific health and safety matters, been given / arranged, before work starts? (This must include particular risks associated with the site) [REG.24]

☐   ☐   ☐ ▶

13. Does the Contractor have details of, and understand, the emergency alarms, evacuation procedures and the use of the emergency equipment and services? [REGS.13 (7) & 22 (1)]

☐   ☐   ☐ ▶

14. Has the Principal Contractor provided work and rest rooms that are suitable, clean and properly maintained? (i.e. with good room heating, ventilation and facilities) [REGS.13(7) & 22(1)]

☐   ☐   ☐ ▶

15. Are the welfare facilities clean, hygienic and properly maintained? (i.e. meeting the minimum regulatory requirements) [REGS.13 (7) & 22 (1)]

☐   ☐   ☐ ▶

General protection [REGS. 22 (1) & 26]	N/A	NO	YES	Details / Comments (in general terms)

16. Are there adequate and effective means of keeping the area/s where the contractor will be working, free from:
- other tradesmen and any hazards arising from their work

☐   ☐   ☐ ▶

- moving plant and vehicles

☐   ☐   ☐ ▶

- persons using nearby site access routes

☐   ☐   ☐ ▶

- members of the public and/or visitors

☐   ☐   ☐ ▶

Note: This can be achieved by physical distance, protective measures to ensure separation (such as a screen), or programming (to separate an adjacent activity in "time")

Other site safety issues [REG.22 (1) (i) (ii)]
17. Are there any other site safety issues that may affect the work?
(These should be listed here or on an attached sheet.)

Any other comments

Principal contractor's directions?
18. Are there any specific directions from the principal contractor? [REG.22(1)] (These should be listed here or on an attached sheet and cross-referenced to the relevant regulation.)

Name: Position:

Signature: Date: Time:

TABLE 6.6 Construction project inspection report form

Name of Project:	Distribution:
	– Site Manager
	– H&S Manager
	– Commercial Manager
Visit Date:	– Project H&S Committee

Part A: SAFETY DOCUMENTATION & COMMUNICATION

		Y/N	Score 1–3★	Comments / Action
1. Safety Notices	• A Notice board clearly displayed for all to see?			
	• Company Safety Policy?			
	• Site Rules displayed?			
	• Appointed safety staff displayed?			
	• Emergency procedures?			
	• Employers' liability cover insurance certificate?			
	• Notification of project to HSE (form F10)?			
2. Construction Phase Plan	• Available on site?			
	• Up to date?			
	• Management structure and responsibilities?			
	• Emergency procedures?			
	• Fire safety plan?			
3. Method statements & risk assessments	• Are safety risk assessments completed?			
	• Are risk assessments completed for project scope?			
	• Are method statements available & approved?			
	• Evidence that method statements have been communicated to personnel?			
4. Site Manager's weekly inspection reports	• Are these available?			
	• Have actions been addressed (closed out)?			
5. Current Report of H&S Manager	• Is this available?			
	• Have actions been addressed (closed out)?			
6. Minutes of the Site safety Committee	• Are these available?			
	• Have actions been addressed?			

7. Records of inductions, plant and permits to work	• Are induction records available and up to date? • Are copies of plant maintenance records available and up to date? • Is a permit to work log/record available and up to date?		
8. Record of Accidents & Incidents	• Have accidents and incidents been recorded appropriately?		

Part B: THE SITE ENVIRONMENT

		Y/N	Score 1–3*	Comments / Action
9. Site security	• Is there a sufficiency of suitable fencing and security on the site to prevent unauthorised access?			
10. Welfare facilities	• Are suitable and sufficient welfare facilities provided in accordance with CDM 2007? • Are the facilities clean? • Are the facilities regularly inspected & maintained?			
11. Materials storage & housekeeping	• Are materials stored in an organised and safe manner? • Is the site tidy of rubbish? • Are recycling bins provided for waste materials? • Is suitable signage placed appropriately around the site? • Are there fire escape routes, assembly points and fire alarms?			
12. PPE	• Is correct PPE worn for all site activities?			
13. Protecting the public	Is the public protected from: • Site traffic • Falling material • Noise/dust/mud			
14. Pedestrian & traffic routes	• Are there separate vehicle & pedestrian access & egress routes? • Are vehicles and pedestrian routes suitably segregated on site?			
15. Site hazards	Are the risks caused by specific site hazards and work activities being managed with the application of suitable control measures?			

(Continued)

TABLE 6.6 (Continued)

16. Interviews with site personnel	• Are they CSCS card holders? • Have they received a site induction? • Are they aware of and working to an approved method statement? • Do they receive regular tool box talks? • Do they have any health and safety concerns?

Score: 3 – Exceeds requirement (excellent); 2 – Meets requirement (good); 1 – Below requirement (poor)

Howarth and Watson (2008)

In undertaking a review of the site environment, regular safety walk inspections enable a review of workplace activities and procedures and any hazards presented in and around the site.

Summary

This chapter has introduced occupational health and safety management systems and has outlined the essential components of such systems. Advocated benefits and problems associated with occupational health and safety management systems have been highlighted and differing standards and guidance documents have been introduced. Issues associated with developing an OHSMS within an organisation have also been introduced.

The health and safety management system of a construction project has been briefly considered and the essential role of inspection and audit has been highlighted. Finally, some examples of useful documentation have been presented to support systematic construction project health and safety management.

Questions for the reader

Here follow a number of questions related specifically to the information presented within this chapter. Try to attempt each question without reference to the chapter in order to assess how much you have learned. The answers are provided at the end of the book.

Question 1

Identify the benefits associated with the deployment of an OHS management system.

Question 2

The Institution of Occupational Safety and Health (IOSH) outlines a process for the development within an organisation of an occupational health and safety management system. Six typical inputs are identified within this process. Identify these six typical inputs.

Question 3

You have been asked to deliver a brief presentation at the next senior management meeting of your department. The presentation concerns an upcoming H&S audit. The title of the presentation is: 'An Outline of the Key Inspection Components of the Upcoming Health and Safety Self-assessment Audit'.

> *Prepare a concise list of the likely key inspection components of the upcoming H&S self-assessment audit.*

Further reading

European Agency for Safety and Health at Work (2002). *The Use of Occupational Safety and Health Management Systems in the Member States of the European Union: Experiences at Company Level.* Luxembourg. Available at: http://osha.europa.eu/en/publications/reports/307. [Accessed 14 June 2017].

European Agency for Safety and Health at Work (2004). *Systems and Programmes Achieving Better Safety and Health in Construction: Information Report.* Available at: http://osha.europa.eu/en/publications/reports/314. [Accessed 8 May 2010].

Health and Safety Executive (1997). *Successful Health and Safety Management,* HSG65. 2nd ed. Sudbury: HSE Books.

Health and Safety Executive (1998). *Managing Health and Safety: Five Steps to Success, INDG275.* Sudbury: HSE Books, 1998 (reprinted 2008). Available at: www.hse.gov.uk/pubns/indg275.pdf.

Health and Safety Executive (2007). *Development of Working Model of How Human Factors, Safety Management Systems and Wider Organisational Issues Fit Together.* Research report RR 543:2007 prepared by White Queen Safety Strategies and Environmental Resources Management for HSE London. Available at: www.hse.gov.uk/research/rrpdf/rr543.pdf.

Health and Safety Executive (2010). *You can assess your own health and safety management system.* Available at www.hse.gov.uk/waste/health-and-safety-management-system-checkilst.doc Sourced 6 Jan 2017.

Laddychuck, S. (2008). Paving the way for world-class performance. *Journal of Safety Research* (39), pp. 143–149. A case study example of the development and implementation of an integrated health, safety and environmental management system within a large business organisation.

Rowlinson, S. (Ed.) (2004). *Construction Safety Management Systems.* London: Spon Press. A compendium of research papers concerning construction safety management systems.

Seiji, M. (2001). *Guidelines on Occupational Safety and Health Management Systems (ILO-OSH 2001) Safe Work.* International Labour Office ILO Geneve.

References

Baker, P. (2001). Raise the safety standard: Horton Kirby. *Works Management*, 54 (11), pp. 26–28.

European Agency for Safety and Health at Work (2002). *The Use of Occupational Safety and Health Management Systems in the Member States of the European Union: Experiences at Company Level*. Luxembourg: Office for Official Publications of the European Communities.

European Agency for Safety and Health at Work (2009). *Working Environment Information Paper: Occupational Safety and Health and Economic Performance in Small and Medium-Sized Enterprises: A Review*. Luxembourg: Office for Official Publications of the European Communities.

Guide to Achieving Effective Occupational Health and Safety Performance. BS 18004: 2008. London: British Standards Institution, 2008.

Howarth, T., and Watson, P. (2008). *Construction Safety Management*. Oxford: Wiley Blackwell.

HSE (2008). *PABIAC Strategic Objective 2: Safety Management Systems. A Self Assessment Tool for SMEs*. Available at: http://www.paper.org.uk/services/health_safety/pabiac/PABIACSafetyManagementSelfAssessmentTool.pdf. [Accessed 21 July 2017].

Institution of Occupational Safety and Health (2009). *Systems in Focus: Guidance on Occupational Safety and Health Management Systems*. Available at: www.iosh.co.uk/techguide. [Accessed 8 May 2010].

Laman, S. (October 2009). Building a consensus. *Quality Progress Magazine*. [Online]. Available at: www.asq.org/quality-progress/2009/10/one-good-idea/building-a-consensus.htm. [Accessed 8 May 2010].

Occupational Health and Safety Management Systems: Guidelines for the Implementation of OHSAS 18001: 2007, OHSAS 18002:2008. London: British Standards Institution, 2008.

Occupational Health and Safety Management Systems Requirements, OHSAS 18001: 2007. London: British Standards Institution, 2007.

Safe Site Access Certificate. (2010). *The Specialist Engineering Contractors Group*. Available at: www.secgroup.org.uk/health.html. [Accessed 1 July 2010].

Street, H. (2000). Getting full value from auditing and metrics. *Occupational Hazards*, 62 (8), pp. 33–35.

Warren, J. (May 2005). *The Role of Performance Measurement in Economic Development*. AngelouEconomics. Available at: www.angeloueconomics.com/measuring_ed.html.

7

BIM AS A QUALITY SYSTEM

Introduction

Building Information Modelling (BIM) is arguably the hottest topic in the field of Construction Management and attracts not only interest, but also controversy. Is it primarily a technology or a process? Does it just apply to building? And what does the 'M' stand for: 'modelling' or 'management'? In fact, given that BIM applies to assets other than buildings, should it be called BIM at all? Here, and for the particular purposes of this book, BIM is treated as a technology-enabled quality system for improving the competitiveness of all organisations operating within the extended construction and property sectors that relates to the whole life of all built assets. The aim of the current chapter is to consider BIM as a quality system and discuss its current state of development in the UK, always bearing in mind the fact that we are dealing with a rapidly moving phenomenon, and one that is global.

Learning outcomes

By the end of this chapter, the reader will be able to demonstrate understanding of:

- The background and key concepts of BIM and its aspects and applications
- The concept of BIM as a process and, in particular, as a quality system
- The advantages and key 'drivers' of BIM adoption
- The barriers to BIM adoption and challenges that need to be overcome
- The ongoing development and future direction of BIM

Background to BIM

Definitions and concepts

Competing definitions of BIM

There are many definitions of BIM and explanations of what it represents. One of the first uses of the term 'building information model' was in a journal article by Van Nederveen and Tolman (1992), and an early appearance of 'building information modeling' came in a Building Industry Solutions White Paper produced by the software company Autodesk (2002). A definition from the (United States) National Institute of Building Sciences (2007) states:

> Building Information Modelling (BIM) is a digital representation of physical and functional characteristics of a facility.

This definition is not ideal: it is actually describing the *model* itself, rather than the process of *modelling*, and in this book we are very much concerned with BIM as a process. However, it does contain two important details. First, a recognition that BIM is not only about 3D geometry but includes 'functional characteristics' (sometimes referred to as 'non-geometric information'); and secondly that the information relates to any 'facility'; not just buildings. In terms of BIM as a process, the latest definition from National Building Specification that BIM is "a process for creating and managing information on a construction project across the project life cycle" (NBS 2016) carries too much of an implication that BIM is for new projects. It isn't: there is massive potential for incorporating BIM into the management of existing built assets as well as using it to create them. This growing awareness of what BIM can or could mean has caused some to question the applicability of the term *Building Information Modelling*. Variants such as *Building Information Management*, *Virtual Design and Construction* and *Digital Design and Construction* have appeared, but only the first avoids the notion that BIM stops once a facility is constructed.

Although it is arguable that BIM (taken literally as *Building Information Modelling*) is far from ideal, the fact remains that it is now commonly accepted and all-pervasive. Of course, outside the English-speaking world, this simply doesn't matter. The expression 'BIM' can embrace all of these meanings without any need for justification (see *Figure 7.1* for an example).

What is involved in BIM?

Having emphasised the idea of BIM as a process, and settled upon the word 'BIM' as it stands, the next step is to go deeper into what exactly is involved. We will do that by looking at the elements of the name itself.

FIGURE 7.1   BIM is global

B is for 'building'

The use of digital technology to support industrial processes did not originate in the Building Industry; other industries, such as Aerospace and Automotive, were earlier adopters. Neither is the potential of BIM limited to creating the assets we call 'buildings', or the sector we call 'Building'. As noted in New Civil Engineer 'the core principles and workflows associated with BIM apply equally to all infrastructure projects' (Corke 2012) and the increasing number of BIM interest groups (representing, for example, clients, retailers, water engineers, health providers, facilities managers and city planners) demonstrates the pervasive nature of BIM.

I is for 'information'

BIM is about the "sharing, analysis and re-use of information" (RICS 2013) and to access information we need to make sense of data. In the case of BIM the data are held digitally in the database (or set of databases) that is the model. As mentioned earlier, the data (and the information that it represents) can be geometric (for example, the dimensions of a window) and non-geometric (e.g. the materials, manufacturer or model number of the window). The information sources from which model data are derived as inputs and the information uses to which the model data can be put, as outputs, are numerous.

M is for 'modelling, or management'

Once again, this is a rather restrictive term. It emphasises the creation of the model, rather than its use, which is the whole point of its existence. Consequently many people prefer the word 'management' rather than 'modelling'.

Drivers for BIM adoption – 'pull' and 'push'

Innovations are adopted in a variety of ways and for a variety of reasons. There are inevitable barriers, and for new ways of working to be adopted, these must be overcome. In economic theory, an innovation might be adopted because of a push by the supply side (i.e. the industry itself) or a pull from the demand side. The UK 'BIM Strategy Report' to the UK Government (BIM Working Party 2011) recommended a combination of both approaches.

The government 'pull'

Within the 'Construction Strategy' published by the UK Government (Cabinet Office 2011) was a requirement for 'collaborative 3D BIM' by 2016. The two key objectives were "cost reduction in the construction and operation of the built environment" and the "implementation of . . . policy in relation to sustainability and carbon". As the public sector in the UK represents a significant portion of construction demand, this 'BIM mandate' could not be ignored by the industry.

The 'push' from the supply side

The criteria for an effective push by the supply side are a business case for adoption and the prospect of a return on investment (ROI). The earliest indications of BIM ROI are from the USA, where Holness (2006), for example, reports savings on construction costs of between 15 percent and 40 percent. In a survey of over 1,000 industry participants in 2012, McGraw-Hill indicated that "63% of BIM users are experiencing a positive perceived ROI on their overall investment in BIM . . . with the most common range being between 10 and 25%" (McGraw-Hill 2012: p. 39). Evidence from the UK is still relatively scarce but suggests that experienced users derive an increased ROI: a realistic suggestion being that BIM adoption may show an initial productivity loss, followed by expected gains later.

The results to date

The Government's 'BIM edict' has produced an enormous interest in BIM in the UK, as illustrated by the annual (since 2011) NBS National BIM Reports (2011–2016). In NBS's 2011 survey (National Building Specification 2011) 43 percent of respondents were unaware of BIM, and this has reduced to 4 percent in the 2016 survey. There was a corresponding increase in those "aware of and using BIM": 13 percent in the 2011 survey, to 54 percent in the 2016 report (National Building Specification 2016).

BIM applications and uses

From its initial use as a 3D enhanced design tool, BIM has extended into a wide range of applications through the project life cycle, some of which have attracted the use of the word 'dimension' as a descriptor: for example '4D BIM' involves linking the 3D model with a time schedule. The following is a reasonably full (though not exhaustive) list of how BIM-based applications can enhance the operations involved in the design, construction and management of built assets.

3D Design and visualization

The earliest application of BIM was in design – architectural, structural and engineering. The design process – iterative in its nature – is made more efficient by the ability to re-use BIM information. Links are available to Structural and Environmental Analysis software, and visualised renderings can be produced for clients and other stakeholders. Because different aspects of design can be integrated within the model, so-called design clashes can be identified and resolved at an early stage, reducing the need for more costly changes on site.

Off-site fabrication

The digital nature of the design data permits the transmission of designs into the automated software-to-manufacturing systems used by some component manufacturers (e.g. structural steel, precast concrete, ductwork), thereby increasing the potential for more efficient off-site manufacture. Referred to as Design for Manufacture and Assembly (DFMA), this brings the benefits of factory production to construction projects, with potential from improvements in health and safety, productivity and the reduction of waste.

Construction planning (4D BIM)

'4D BIM' involves linking a time schedule to a 3D model to improve the time planning and control of construction activities. Schedules can be generated by interrogating the design model(s) to identify activities, calculate durations (using automated quantity extraction), impose assembly and installation logic, schedule resource requirements and visualise construction through animations of the process.

Commercial management (5D BIM)

'5D BIM' caters for estimating, cost management and procurement. This includes 'time-cost-value' analysis techniques such as Earned Value Management (see e.g. Barlish and Sullivan 2012: p. 153). Work is also under way to integrate BIM applications with enterprise resource planning (ERP) systems at the business level of the organisation (see e.g. Babič et al. 2010) to inform their sales, purchasing and logistics functions.

Sustainability (6D BIM)

BIM for Sustainability ('6D BIM') allows information such as energy use, resource efficiency and other aspects of sustainability to be better analysed, managed and understood (see e.g. Hamza and Horne 2007; Azhar and Brown 2009; Nour et al. 2012). The BIM model can accommodate information such as embodied carbon, including that created by the process of construction.

Facilities management (7D BIM)

7D BIM tackles the management of facilities or assets. An 'asset tagged' BIM model, delivered to the client or end-user on completion can be populated with appropriate component and product information, operation manuals, warranty data and so on. Information based on BIM can thus be re-used for driving efficiencies in the management, renovation, space planning and maintenance of facilities. This potential applies not just to the handover of new built assets but to the retrospective modelling of existing ones through point cloud capture using laser scanning (see Volk et al. 2014).

Aspects of BIM: technology, process and people

A common approach to examining any production system (including the impact of innovation on it) is through the three aspects of *technology, process* and *people*. Most discussion is over what happens when one of the three intervenes to cause a change in the relationship between the other two (e.g. the impact of a new *technology* on *people* and *process*). Furthermore, as US architect John Tobin observed, BIM was initially used as "a sustaining technology"; as a 3D tool that uses models to produce construction designs more efficiently: it has moved on to be a "disruptive technology", one that invites the re-imagining of the whole process, allowing us to "change markets and expectations" (Tobin 2013). The main focus of this chapter is on BIM as a process, but the other aspects require a few words before considering this in detail.

BIM as technology

BIM is supported by a variety of software platforms, usually proprietary and sold commercially. Eastman's original *BIM Handbook* (2008) lists over 70 different software companies with hundreds of different software packages. These have been developed to suit the functional needs of their target users (architects, structural engineers, services engineers, constructors . . .) and thus differ structurally and semantically (see Lockley et al. 2013). In the 2014 NBS National BIM Report, 25 percent of BIM users reported that "information models only work in the software they were made on" (NBS 2014: p. 14). This is referred to as an issue of 'interoperability'. It is a limiting factor in achieving fully collaborative BIM, and various organisations, including the International Organization for Standardization (ISO), have worked to improve the reliable exchange of data between 'native' software platforms. The result is an intermediary format called *Industry Foundation Classes* (IFC). The latest version, IFC4, is registered as ISO 16739:2013. Recognising the importance of IFC standards, most producers of BIM software platforms now aim to ensure that their products support them fully and are, through IFC, compatible with one another.

BIM and people

For any innovation to be successful, people must be willing to adopt it, and in many cases this requires a commitment to acquiring new skills and knowledge. Evidence from the 2016 National BIM Report shows that although 27 percent of those yet to adopt BIM would 'rather not', most industry professionals are positive towards BIM, with only 6 percent of those who have adopted BIM saying that they wished they hadn't (NBS 2016: p. 37).

BIM as a process

In the foreword to their *BIM Handbook*, Eastman and his co-authors note that BIM adoption requires not just a change in technology but a 'process change' (Eastman et al. 2011). Reflecting this, the 2016 NBS National BIM Report shows that 92 percent of BIM users agreed that "adopting BIM requires changes in our workflow, practices and procedures" (National Building Specification 2016: p. 37). It is precisely this aspect of BIM that accounts for its identification as a 'disruptive innovation', and we will explore what this means in the following sections.

Collaborative production and use of information

The UK Government's 'BIM mandate' was for 'collaborative 3D BIM'. In other words, to exploit the full potential of BIM, collaboration is required. An earlier drive towards collaborative design was exemplified in the British Standards Institute's 'Code of Practice for Collaborative Production of Architectural, Engineering and Construction Information' (BS 1192:2007) but in the BIM era this collaboration has been extended beyond the production of information to its *use*. A key component is the concept of a Common Data Environment (CDE) – "a single source of information . . . for multi-disciplinary teams in a managed process" (BSI 2013). CDEs are not limited to digital data, but most of the providers of file-sharing project extranets offer products that fulfil the function with BIM databases, rather than files.

Innovative approaches to project procurement

It could be argued that traditional approaches to the way projects are procured compromise the ability to collaborate right from the start. The American Institute of Architects (AIA) have advocated a project delivery approach called 'Integrated Project Delivery' (IPD) that "integrates people, systems, business structures and practices into a [collaborative] process . . . through all phases of design, fabrication and construction" (AIA California Council 2007).

In an initiative that runs parallel to its commitment to BIM, the UK Government Cabinet Office has proposed three 'new models of procurement' that would best correspond to "high levels of supply chain integration, innovation, and good working relationships"(Cabinet Office 2014: p. 7). These are:

> *Cost-led procurement*: The most conventional of the three 'new models' in which an 'integrated framework supply team' is selected from up to three competing bids, based on affordability and quality criteria.
>
> *Two-stage open book*: A first stage, in which contractor-consultant teams compete on the basis of a development fee and qualitative elements is followed by a second, where the successful team openly develop the project proposal to the client's cost benchmark.
>
> *Integrated Project Insurance*: Following competition based upon qualitative criteria and a 'fee declaration', an integrated project team is selected to develop an acceptable design solution and a single joint-names project insurance policy is executed to cover all risks associated with delivery of the project.

BIM procedures and protocols

The increased collaboration that is required for more effective BIM exploitation brings a variety of accompanying challenges. For projects to achieve 'Level 2[1] BIM' (i.e. meet the requirements of the UK Government BIM mandate) it is necessary to set rules, conventions and ways of working to cope with the individuality of the different project participants. There are technical issues around data exchange and interoperability, as discussed earlier. But when BIM becomes collaborative, there are also questions, up and down the project supply chain, as to *what* information is to be expected *when*, *from whom*, *to whom* and in *what form* or level of development. Without some form of standardisation, there would be the prospect of chaos, and the UK Government has taken a lead in creating standardised solutions to some of these questions.

Earlier, a number of standard protocols for the BIM process had been developed in the United States, most notably the AIA's 'Digital Practice Documents' (AIA 2007), and in these the key requirements for a BIM-enabled project were identified. These include:

- *Employer's Information Requirements (EIR)*, which sets out a client's requirements for the delivery of information by its project supply chain.
- *BIM Execution Plan (BEP)* or *Project Execution Plan (PEP)*, which demonstrate how the EIR will be delivered and which can contain a *Master Information Delivery Plan* (MIDP) to indicate when project information is to be produced, by whom and how.
- The *Project Information Model (PIM)*, consisting of all the documentation, non-graphical and graphical information that defines the delivered project and which (for the purpose of managing, maintaining and operating the asset) is eventually superseded by the *Asset Information Model* (AIM).
- A *Common Data Environment (CDE)* that, amongst other things, will contain all of the above.

Standardised process solutions and systems

Since 2011 the UK Government has commissioned 'standardised solutions' for working in BIM. Currently, these take the form of nine components – official standards, implementation tools and guides that comprise the 'rules of engagement' for Level 2 BIM. Some are designated as 'British Standard' (BS) whilst others are named 'Publicly Available Specification' (PAS[2]). They are:

- *PAS 1192–2:2013 (BSI 2013) Specification for information management for the capital/delivery phase of assets using building information modelling*: This document builds upon the existing BS 1192:2007 (see above) by specifying what is required for delivering projects in Level 2 BIM and describes how models evolve through increasing levels of development.
- *PAS 1192–3:2014 (BSI 2014) Specification for information management for the operational phase of assets using building information modelling*: PAS 1192–3 extends the project information delivery cycle into the operating phase of the built asset's life cycle. It specifies information requirements from the viewpoint of the operational phase of a constructed asset or group of assets.
- *BS 1192–4 Collaborative production of information. Part 4: Fulfilling employers' information exchange requirements using COBie*: This represents a revision of BS 1192:2007 to encompass the handling of information using COBie. COBie is a data schema presented in the form of a spreadsheet which serves as a standardised index of information about new and existing assets throughout their life cycle.
- *PAS 1192–5:2015 (BSI 2015) Specification for security-minded building information modelling, digital built environment and smart asset management*: The document provides stakeholders with protocols and controls to ensure the security of their data whilst they are collaborating digitally.
- *PAS 1192–6:2015 (draft) Specification for collaborative sharing and use of structured hazard and risk information for health and safety*: The aim of this draft PAS is to specify the H&S requirements that should be 'embedded into all BIM projects' at their outset.
- *CIC BIM Protocol (CIC 2013)*: This is a document, for projects in a BIM environment, to supplement the standard forms of contract that parties use.
- *Government Soft Landings (GSL)*: GSL is a protocol that specifies the handover of an asset (with its information) to assist owners and their asset managers.
- *BIM Toolkit and Digital Plan of Work/Digital Toolkit*: This a customisable digital delivery template for project information. Set against the eight stages of the RIBA's Plan of Work (RIBA 2013) it allows users to specify and verify the delivery of geometric and other data and documentation to the client of required levels of development. It defines and allocates responsibilities for this and assists in verification of information delivery.
- *Unified BIM Classification System*: Standard classification systems are an essential part of information management and data sharing, as they provide the logic to enable the search for and retrieval of information, and its integration and aggregation.

BIM prospects, barriers and future

The current state of BIM adoption and the immediate future

The 2016 'deadline' set by the UK Government has now passed. The most recent available information on how the industry has responded is to be found in NBS's 6th annual BIM Report (NBS 2016). It reveals that a majority (54 percent) of the 1,000+ construction industry professionals who responded are 'aware of and using BIM' on some of their projects and 86 percent expect to be doing so by 1 year's time. Just under half (49 percent) feel 'confident of their BIM knowledge and skills'. Of those, 70 percent use BIM (they produce 3D models) and of these, 56 percent have shared models with external designers and 45 percent with parties across different disciplines. Most use of BIM appears to be still restricted to the design stages (only 37 percent use models from the start to end of a project) and only 16 percent pass on a model to FM). Almost all (90 percent) respondents feel that BIM requires (or would require, in the case of non-adopters) changes in their workflow. In terms of the influence of BIM on the 2011 Government Construction Strategy targets: 63 percent believe a 33 percent reduction in cost is possible; 57 percent a 50 percent reduction in time; 39 percent a 50 percent reduction in BE greenhouse gas emissions; less than 1/3rd that BIM will help create a trade gap reduction.

Barriers and challenges

In an earlier part of the chapter, some of the problems with operating collaboratively in BIM were identified. As shown in the previous sections, guidance has been forthcoming and continues to emerge from government sources, professional bodies from software developers. Matters such as computer failure and data security will have increased importance. Finally, as shown in studies, such as those by van der Smagt (2000) and Dossick and Neff (2010), 'human factors' (such as leadership, capability, education, organisational culture, teamworking) play a leading part in the likely success of BIM-enabled construction operations.

The future of BIM

A strategy for the future of BIM entitled 'Digital Built Britain' was published in October 2014. It previews the extension of the digital revolution to the "the way we plan, build, maintain and use our social and economic infrastructure". The predicted technological adoptions that flow from this digital revolution will include *predictive digital decisions*; i.e. solutions based upon digital information which in some cases can be automated, that is, require no human intervention. A month earlier, the 'BIM2050 Group' produced its own report on 'Our Digital Future' (CIC BIM2050 Group 2014). The summarised findings include:

* *Cyber Security*: the conflict between free and open information, connectivity and collaboration, on the one hand, and the need for information security;

- *Interoperability for Smart Cities*: which recognises the advent of 'intelligent infra-structure' in future smart cities; and
- *Nano-second Procurement and Performance*: the ability of more businesses to approach the reaction speeds that are currently in existence in the stock mar-ket through digitisation of their business management and enterprise resource planning systems.

Summary

It is appropriate to finish this chapter with an extract from the Executive Summary of the 'Digital Built Britain' publication.

> Building Information Modelling (BIM) is changing the UK construction industry – a vitally important sector that employs more than three million people and in 2010 delivered £107 billion to the UK economy. Over the next decade this technology will combine with the internet of things (pro-viding sensors and other information), advanced data analytics and the digital economy to enable us to plan new infrastructure more effectively, build it at lower cost and operate and maintain it more efficiently. Above all, it will enable citizens to make better use of the infrastructure we already have.
>
> *(H.M. Government 2014: p. 5)*

Questions for the reader

Here follows a number of questions related specifically to the information presented within this chapter. Try to attempt each question without reference to the chapter in order to assess how much you have learned. The answers are provided at the end of the book.

Question 1

What are '5D' and '6D' BIM?

Question 2

The key requirements for a BIM-enabled project have been identified as including?

Question 3

What is COBie ?

Question 4

The UK Government Cabinet Office has proposed three 'new models of pro-curement' that would best correspond to 'high levels of supply chain integration, innovation, and good working relationships'. Identify these three 'new models of procurement'.

Notes

1   A fuller description of the different levels of BIM maturity is provided in Chapter 8.
2   A PAS is a less formal version of a BS, but similarly structured and usually developed to meet an urgent need.

Further reading

BIM Task Group (2013). *Government Soft Landings.* Available at: www.bimtaskgroup.org/gsl/ [Accessed 24 March 2017].

Gu, N., and London, K. (2010). Understanding and facilitating BIM adoption in the AEC industry. *Automation in Construction,* 19 (8), pp. 988–999.

References

AIA California Council (2007). *Integrated Project Delivery: A Working Definition (Version 2).* Sacramento, CA: McGraw-Hill Construction and American Institute of Architects, California Council.

Autodesk (2002). *Building Information Modeling.* Autodesk Building Industry Solutions. Available at: www.laiserin.com/features/bim/autodesk_bim.pdf. [Accessed 24 March 2017].

Azhar, S., and Brown, J. (2009). BIM for sustainability analyses. *International Journal of Construction Education and Research,* 5 (4), pp. 276–292.

Babič, N.C., Podbreznik, P., and Rebolj, D. (2010). Integrating resource production and construction using BIM. *Automation in Construction,* 19, pp. 539–543.

Barlish, K., and Sullivan, K. (2012). How to measure the benefits of BIM: A case study approach. *Automation in Construction,* 24, pp. 149–159.

BIM Industry Working Group (2011). BIM Management for value, cost and carbon improvement. Strategy paper for the Government Construction Client Group. Available online at: http://www.bimtaskgroup.org/wp-content/uploads/2012/03/BIS-BIM-strategy-Report. pdf. [Accessed 21 July 2017].

BIM Task Group (2013). *Building Information Model (BIM) Protocol.* Construction Industry Council, London.

BSI (2013). *PAS 1192–2 Specification for Information Management for the Capital/Delivery Phase of Construction Projects Using Building Information Modelling.* London: BSI Standards Limited.

BSI (2014). *PAS 1192–3 Specification for Information Management for the Operational Phase of Assets Using Building Information Modelling (BIM).* London: BSI Standards Limited.

BSI (2015). *PAS 1192-5 Specification for Security-Minded Building Information Modelling, Digital Environments and Smart Asset Management.* London: BSI Standards Limited. London.

Cabinet Office (2011). *Government Construction Strategy.* Available at: www.gov.uk/government/ uploads/system/uploads/attachment_data/file/61152/Government-Construction-Strategy_0.pdf. [Accessed 24 March 2017].

Cabinet Office (2014). *New Models of Construction Procurement.* Available at: www.gov.uk/ government/publications/new-models-of-construction-procurement-introduction. [Accessed 24 March 2017].

CIC BIM2050 Group (2014). *Built Environment 2050: A Report on Our Digital Future.* Construction Industry Council BIM2050 Group. Available at: www.bimtaskgroup.org/wp-content/uploads/2014/09/2050-Report.pdf. [Accessed 24 March 2017].

Corke, G. (2012). *BIM: Constructing a Virtual World.* New Civil Engineer. Available at: www. nce.co.uk/bim-constructing-a-virtual-world/8640433.article. [Accessed 24 March 2017].

Dossick, C.S., and Neff, G. (2010). *Messy Talk and Clean Technology: Requirements for Interorganizational Collaboration and BIM Implementation within the AEC Industry.* In Taylor, J.E., and Chinowsky, P. (Eds.). Proceedings Editors. South Lake Tahoe, CA.

Eastman, C. (2008). *BIM Handbook, A Guide to Building Information Modeling for Owners, Managers, Designers, Engineers, and Contractors.* Hoboken, NJ: John Wiley & Sons.

Eastman, C., Teicholz, P., Sacks, R., and Liston, K. (2011). *BIM Handbook: A Guide to Building Information Modeling for Owners, Managers, Designers, Engineers and Contractors.* 2nd ed. Hoboken, NJ: John Wiley and Sons.

Hamza, N., and Horne, M. (2007). *Building Information Modelling: Empowering Energy Conscious Design.* 3rd International Conference of the Arab Society for Computer Aided Architectural Design (ASCAAD), November 26–28, Alexandria, Egypt.

Holness, G. (2006). Building Information Modeling. *Ashrae Journal,* 48 (8), pp. 38–40, 42, 44–46.

H.M. Government (2012). *Industrial Strategy: Government and Industry in Partnership.* Available at: www.gov.uk/government/uploads/system/uploads/attachment_data/file/34710/12-1327-building-information-modelling.pdf. [Accessed 24 March 2017].

H.M. Government (2014). *Digital Built Britain.* Available at: www.digital-builtbritain. comDigitalBuiltBritainLevel3BuildingInformationModellingStrategicPlan.pdf. [Accessed 24 March 2017].

Lockley, S., Greenwood, D., Matthews, J., and Benghi, C. (2013). Constraints in authoring BIM components for optimal data reuse and interoperability: Results of some initial tests. *International Journal of 3-D Information Modeling (IJ3DIM),* 2 (1), pp. 29–44.

McGraw-Hill Construction (2012). *Business Value of BIM for Construction in North America.* Available at: http://static-dc.autodesk.net/content/dam/autodesk/www/campaigns/BTT-RU/MHC-Business-Value-of-BIM-in-North-America-2007–2012-SMR.pdf. [Accessed 24 March 2017].

National Building Specification (2011–2016). *NBS National BIM Report 2011–2016.* Available at: www.thenbs.com/topics/bim/reports/index.asp. [Accessed 24 March 2017].

National Institute of Building Sciences (2007). *National BIM Standard-United States™.* Available at: www.nationalbimstandard.org/faq.php#faq1. [Accessed 24 March 2017].

Nour, M., Hosny, O., and Elhakeem, A. (2012). A BIM based energy and life cycle cost analysis/optimization approach. *International Journal of Engineering Research and Applications,* 2 (6), pp. 411–418.

RIBA (2013). Plan of Work. Royal Institute of British Architects. London.

RICS (2013). BIM Survey Results. The Royal Institute of Chartered Surveyors. London.

Tobin, J. (2013). *BIM Becomes VDC: A Case Study in Disruption.* Building Design and Construction Network. Available at: www.bdcnetwork.com/bim-becomes-vdc. [Accessed 31 May 2015].

van der Smagt, T. (2000). Enhancing virtual teams: Social relations v. communication technology. *Industrial Management & Data Systems,* 100 (4), pp. 148–156.

Van Nederveen, G.A., and Tolman, F.P. (1992). Modelling multiple views on buildings. *Automation in Construction,* 1 (3), pp. 215–224.

Volk, R., Stengel, J., and Schultmann, F. (2014). Building Information Modeling (BIM) for existing buildings: Literature review and future needs. *Automation in Construction,* 38, pp. 109–127.

8

ASSESSING AND DEMONSTRATING BIM CAPABILITY

Introduction

In the previous chapter BIM was introduced. The emphasis was on BIM as a quality system in itself, the benefits BIM might bring and why organisations within the sector are keen to enjoy its benefits.

Different organisations will adapt BIM differently and at different rates of progress. How can a particular organisation's BIM maturity be demonstrated to clients, stakeholders and the world at large? This is the core issue that this chapter addresses – by examining how the quality of an organisation's BIM processes can be demonstrated.

Starting with an outline of what criteria must be met and how they can be measured, this chapter gives consideration to the emerging number of BIM certifying schemes and what it is they are purporting to certify. Finally, a view is presented of what the future is likely to hold for those wishing to acquire, confirm and demonstrate their 'BIM credentials'.

Learning outcomes

By the end of this chapter the reader will be able to demonstrate an understanding of:

- Different levels of BIM maturity and how they are described
- The components of Level 2 BIM and the key characteristic of projects operating at Level 2
- The requirements of PAS91:2013 as an example of Level 2 BIM compliance
- The current options available to an organisation to demonstrate Level 2 BIM compliance (e.g. in the form of currently available BIM certification schemes)
- The availability of certification (by an accredited third party) for demonstrating Level 2 BIM compliance
- Likely future developments in BIM certification (e.g. by the International Organization for Standardization).

BIM maturity

Defining BIM maturity

The Bew Richards Model

A measure of the extent of BIM adoption is provided by the 'BIM Maturity Diagram' developed by Mervyn Richards and Mark Bew in 2008 (*see Figure 8.1*), and subsequently adopted by the UK Government's BIM Task Group (see www.bimtask group.org/).

A similar representation has been produced by Succar (2009); however, it is the Richards-Bew diagram and its terminology that have become most widely accepted. In the diagram there are four BIM maturity levels (from 0 to 3). The starting point – 'Level 0' – represents a situation where designers are working using manual methods or CAD, and the outputs are passed on for construction as documents. 'Level 1' sees the introduction of 3D modelling, though this is isolated (sometimes referred to as 'lonely BIM') with no collaboration between disciplines. In Level 2 there is some degree of model collaboration. Typically, this would involve 'federating' individual models to work in a 'common data environment'. It is 'Level 2 BIM' that is of current interest, the reasons being that

- 'Level 2 BIM' was the target of the UK Government's 'BIM Mandate' and represents what the industry is working to achieve;
- it is now reasonably well established what 'Level 2 BIM' entails; and
- there are standard documents to assist organisations reaching BIM Level 2.

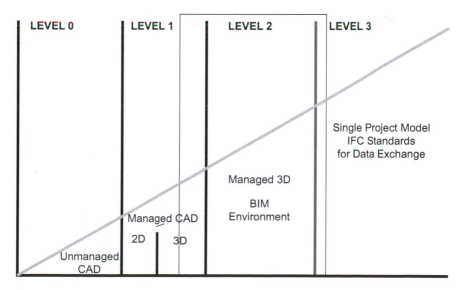

FIGURE 8.1 BIM maturity diagram
(developed by Richards and Bew)

The key characteristic of projects operating at Level 2 is some degree of model collaboration and sharing, typically by 'federating' individual models to work in a 'common data environment' that allows the sharing of digital design models and supporting information.

The final stage of the Richards-Bew diagram, Level 3, is relatively undefined and currently aspirational. However, a view of what BIM Level 3 might be like is given in the Government's 2014 publication 'Digital Built Britain: Level 3 Building Information Modelling Strategic Plan' (H.M. Government 2014).

Measuring BIM maturity

It is one thing to describe the maturity levels. Actually measuring whether they have been achieved is quite another thing. There have been several attempts to create evaluation tools, the earliest being in the United States. In 2007, the US National Institute of Building Sciences developed a Capability Maturity Model with 11 categories and five ratings (Minimum BIM, Certified, Silver, Gold, Platinum) (NIBS 2007). In 2009 the BIM proficiency matrix was developed at Indiana University and assesses BIM proficiency in 8 categories. 'BIM QuickScan' is an evaluation tool widely used in the Netherlands from its development in 2009 and consists of multiple-choice questions over four areas of Management, Culture, Information Structure and Tools and Applications (Berlo et al. 2012). The 'VDC Scorecard' was proposed and tested by academics at Stanford University. It assesses the VDC (effectively, BIM) maturity of a project "across 4 Areas, 10 Divisions, and 56 Measures, and deploys the Confidence Level measured by 7 factors to indicate the accuracy of scores" (Kam et al. 2013). Work by Succar and colleagues has created a number of flexible assessment tools having first identified individual BIM competencies as "the building blocks of organisational capability" (Succar et al. 2013). In 2017, a useful 'Overview of BIM Maturity Measurement Tools' was published by Wu et al. (2017).

Demonstrating BIM competence

Why the need to demonstrate BIM capability?

There are internal and external benefits for any organisation in being able to demonstrate the quality of its people, products, services, processes or systems. Ideally this is demonstrated through some form of third party certification; better still, it is certification that is nationally or internationally recognised.

From an internal perspective, the efforts involved in demonstrating quality will enhance the performance of the organisation and create a culture of improvement. Externally, it promotes credibility amongst customers: sometimes it is a necessity. In Singapore, for example, certification to the ISO 9000 series has been made compulsory for construction contractors and consultants who want to be registered to carry out public-sector projects of a certain size (Ofori and Gang 2001).

Government mandates around the world

In terms of BIM, the background to the need for organisations to demonstrate their competence lies in the increasing number of public administrations (countries, federal states, municipalities and agencies) that have followed the lead of the UK Government in requiring, or at least encouraging BIM. Public sector BIM standards or requirements in Norway, Denmark, Finland and Sweden pre-date those of the UK but tend to be devolved to organisations such as the 'Statsbygg' government agency in Norway. Also in advance of the UK moves was Singapore, which, from 2015, have required BIM-based 'e-submissions' for development approval of all projects above 5,000 m². The USA, being federal, has no national requirement but stipulations come through the medium of agencies such as the General Services Administration, State Department or Department of Defense. France has implemented a three-year national plan for a 'Transition to digital construction'. From 2016 Italy requires BIM for public works over €5 m, and Russia for all public projects from 2019. Spain is considering a mandate starting sometime in 2018. Some such mandates are selective, for example, in Germany, where the intention is that BIM will be required for federal transport infrastructure by 2020, and in Dubai, which has mandated BIM for Architectural and MEP services for all buildings over 40 storeys, greater than 25,000 m², projects led by an international company and all hospitals, universities and similar buildings.

Such requirements put a clear onus on construction organisations to be able to demonstrate their BIM credentials.

UK requirements to demonstrate BIM capabilities

The most important example, within the UK, of a requirement to demonstrate BIM capability is PAS91. *Publicly Available Specification – PAS91:2013- 'Construction prequalification questionnaires'* is a revision of an earlier document (PAS91:2010) that had been produced to streamline pre-qualification procedures for public sector construction projects. Although it is specifically created for public sector construction it can be (and is) readily adapted for use by private sector clients. It is, therefore, arguably the single most important example of construction pre-qualification requirements in the UK.

PAS91 has a set of 'Core question modules' (including *Financial information, business & professional standing* and *health & safety*) and 'Optional question modules' (to be asked 'when judged relevant') and it is to this set that, in 2013, *Module 04: Building information modelling policy and capability* was added. A prefacing note to the module states: "This will be used for UK Government procured projects for Departments that have commenced implementation of the BIM Strategy and may be used by other clients adopting a similar path" (PAS91:2013: p. 23).

Assessing and demonstrating BIM policy and capability

The criteria for BIM policy and capability compliance that are contained in PAS91 are outlined in a series of questions to which the supplier can answer 'yes' or 'no' and give reference to 'relevant supporting information'. The questions are:

- Q1 Do you have the capability of working with a project using a "Common Data Environment" as described in PAS 1192:2:2013?
- Q2 Do you have documented policy, systems and procedures to achieve "Level 2 BIM" maturity as defined in the government's BIM Strategy?
- Q3 Do you have the capability of developing and delivering or working to (depending upon the role(s) that this PQQ covers) a BIM Execution Plan (BEP) as described in PAS 1192:2:2013?
- Q4 Do you have arrangements for training employees in BIM-related skills and do you assess their capabilities?

The first requirement, understanding of and ability to operate within a Common Data Environment (CDE) includes the ability to demonstrate that an organisation is able to "exchange information between supply chain members in an efficient and collaborative manner": for example, by reference to a project that has already been delivered. A fuller description of the CDE is given in Chapter 7.

The second, relating to documented policy, systems and procedures to achieve Level 2 BIM, should be applicable to all projects the organisation is likely to undertake, whatever their size, authorized by the Chief Executive Officer (or equivalent) and regularly reviewed.

The third requirement bulleted above is that the applicant demonstrates a role-related understanding of the appropriate mechanisms of 1192:2:2013 – e.g. in the case of a main contractor, the provision of a BIM Execution Plan (BEP). Again, descriptions of tools such as BEPs are given in Chapter 7.

Finally, a successful applicant for pre-qualification must show that it has suitable training and assessment arrangements to create 'sufficient skills and understanding' for its workforce to implement and deliver projects at Level 2 BIM maturity. Here, the expression 'its workforce' is likely to be interpreted to include the applicant's supply chain, but the word 'sufficient' indicates a level of reasonableness: for example, a Tier 3 sub-subcontractor would not be expected to know how to prepare a BIM Execution Plan.

Demonstrating compliance with PAS91

For this, and any other part of the PAS, there are three ways that the client can assess (and the supplier demonstrate) compliance with the above criteria:

- *Verification and assessment* would include an onsite audit. This would be the most time- and resource-consuming but would offer the greatest certainty to the client;
- *Validated assessment* is less expensive, being a desktop assessment of relevant paperwork (certificates, procedures, etc.); and
- *Self-assessment* is an option where the supplier makes statements within the questionnaire itself about compliance.

In fact, there is a fourth way to demonstrate compliance with PAS91 Module 04 and that is by 'exemption'. The corresponding note states (PAS91 p. 23): "The questions in this module need not be completed if your organization holds a third party certificate of compliance with BS PAS 1192:2:2013 . . . from an organisation with a related UKAS accreditation, or equivalent". Clearly, holding such a 'certificate of compliance' is preferable to having to demonstrate compliance (in any of the three ways) each and every time an organisation seeks to pre-qualify for a project. It is the availability of certification (by an accredited third party) for demonstrating Level 2 BIM compliance to which we now turn our attention.

Accreditation and certification structure

The concepts of certification of an organisation by a third party that is accredited to do so should now be a familiar one to readers of this book.

To summarise: in the UK, the appointed National Accreditation Body is the United Kingdom Accreditation Service (UKAS). The role of UKAS is to evaluate the capability of third party conformity assessors, who, if accredited, are able to act as independent third-party 'certifiers'.

This structure is illustrated in *Figure 8.2*.

As a note to its exemption statement, *PAS91 Optional Question Module O4: BIM, policy and capability* states: "Such accrediting organizations will be required to have specialized design management competences".

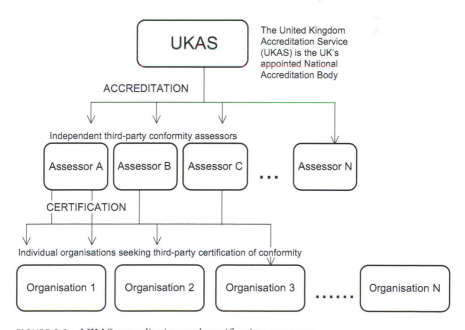

FIGURE 8.2   UKAS accreditation and certification structure

Current BIM certification schemes

The drive for firms in the construction industry to demonstrate BIM competence in the lead-up to the UK Government's 2016 BIM mandate led to a number of organisations offering training and certification in aspects of BIM. This situation has been described by BIM commentator Terry Gough (2015) as a "recipe for confusion".

To recap, the exemption criteria of PAS 91:2013 that relate to BIM policy and capability are:

*   that there is a certificate of compliance with BS PAS 1192:2:2013, and
*   that the certificate of compliance is from a third party organisation with a related UKAS accreditation or equivalent.

A number of organisations are currently offering some form of recognised certification that purports to satisfy these exemption criteria. These include BRE Global (part of the Building Research Establishment), the British Standards Institute (BSI) and LRQA (part of the Lloyd's Register Group). There is a degree of variation in the language of what is offered and in what criteria for BIM compliance are adopted by these organisations, but the basic requirements are derived from PAS 1192–2:2013. LRQA refers to the ability to demonstrate "full compliance with PAS 91:2013, PAS 1192–2:2013, PAS 1192–3:2014, BS11000–1:2010 and all published sections of BS 1192", while BSI have a 12-step route, via a 'BIM verification certificate' to ultimately a 'bsi BIM Design and Construction Kitemark™'. In most cases these organisations offer an integrated system of guidance (e.g. in the form of awareness-making and 'gap analysis') through stages of training to support achievement of certified compliance.

Of course there are many organisations offering BIM training and certification: some may satisfy the requirements of PAS91 but until the creation of a more easily recognised national or international standard that specifically relates to BIM, there remains the possibility of the industry 'confusion' noted earlier and doubts over the credibility of some schemes. It is the prospect of more certain standards that the next section addresses.

Future developments in BIM certification

When PAS 91 exemption statements for optional pre-qualification requirements refer to evidence of third party certificates of compliance, they invariably relate to easily recognised national, or international standards. For example, for Optional Question Modules O1 (Equal opportunity and diversity), O2 (Environmental management) and O3 (Quality management) the reference is to 'compliance with BS EN ISO 9001'. However, in *Optional Question Module O4* the exemption statement for BIM policy and capability refers to a requirement for "third party certificate of compliance with BS PAS 1192:2:2013". BIM policy and capability is the odd one out, as yet having no recognised international standard. However, there is a standard under development under the mantle of the International Organization for Standardization (ISO).

The development of an international standard

The proposed standard, ISO/DIS 19650: *Organization of information about construction works – Information management using building information modelling*, comes in two parts, both of which can be previewed at www.iso.org/standard/68080.html.

Part 1 of the standard is entitled *Concepts and Principles*, and Part 2, *Delivery Phase of Assets* indicating its focus on delivery rather than the subsequent management of the asset (a phase in the life cycle that is already covered by a separate ISO, 55000). Work on ISO/DIS 19650 is not complete but it has built substantially upon PAS 1192–2:2013 and the production and delivery of construction project through a managed process and in a common data environment and using concepts familiar to UK BIM users such as the Project Information Model (PIM) and the Asset Information Model (AIM).

It is envisaged that the new ISO standard will sit within existing ISO series (e.g. ISO 9001 and ISO 55000) as shown in *Figure 8.3*.

There is a need for integration with other existing standards such as ISO 14001 (Environmental Management) and OHSAS 18001/ISO 45001(Occupational Health and Safety Management) to support BIM processes and ISO 44001 (Collaborative business relationship management systems), which has in 2017 replaced BS 11000.

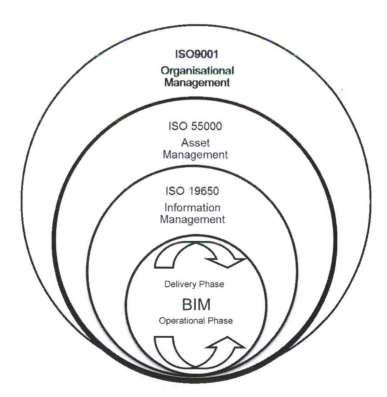

FIGURE 8.3   How the new BIM ISO standard could sit within existing ISO series

Summary

The chapter has examined *why* organisations, particularly supply-side organisations such as design consultants and contractors might wish to demonstrate their BIM capabilities and *how* they can do it. Increasing levels of BIM maturity have been described and the components of the different levels outlined, in particular that labelled 'Level 2 BIM'. The recently inserted BIM criteria within a standard public pre-qualification document, PAS91:2013, have been examined in depth. A way of achieving these criteria – by certification of an accredited third party – has been explored along with other options for certification. A brief review of quality assurance in the UK has been provided and the chapter comes to a close with a look ahead to likely future developments in BIM certification.

Questions for the reader

Here follows a number of questions related specifically to the information presented within this chapter. Try to attempt each question without reference to the chapter in order to assess how much you have learned. The answers are provided at the end of the book.

Question 1

Project suppliers are required to answer BIM policy and capability compliance questions contained in PAS91.

Identify four BIM policy and capability compliance supplier questions that project suppliers are required to answer.

Question 2

What are the exemption criteria of PAS 91:2013 that relate to BIM, policy and capability?

Question 3

Name 3 organisations that offer some form of recognised BIM certification that purports to satisfy these PAS exemption criteria

Further reading

BIM Task Group (2013). *Government Soft Landings*. Available at: www.bimtaskgroup.org/gsl/. [Accessed 03 June 2015].

BIM Working Party (2011). *BIM: Management for Value, Cost and Carbon Improvement: A Report for the Government Construction Client Group*. Available at: www.bimtaskgroup.org/wp-content/uploads/2012/03/BIS-BIM-strategy-Report.pdf. [Accessed 03 June 2015].

BSI (2013). *PAS 1192–2 Specification for Information Management for the Capital/Delivery Phase of Construction Projects Using Building Information Modelling*. UK: BSI Standards Limited.

BSI (2014). *PAS 1192–3 Specification for Information Management for the Operational Phase of Assets Using Building Information Modelling (BIM)*. UK: BSI Standards Limited.

Eastman, C., Teicholz, P., Sacks, R., and Liston, K. (2011). *BIM Handbook: A Guide to Building Information Modeling for Owners, Managers, Designers, Engineers and Contractors*. 2nd ed. Hoboken, NJ: John Wiley and Sons.

H.M. Government (2012). *Industrial Strategy: Government and Industry in Partnership*. Available at: www.gov.uk/government/uploads/system/uploads/attachment_data/file/34710/12-1327-building-information-modelling.pdf. [Accessed 03 May 2015].

van der Smagt, T. (2000). Enhancing virtual teams: Social relations v. communication technology. *Industrial Management & Data Systems*, 100 (4), pp. 148–156.

References

Berlo, L.V., Dijkmans, T., Hendriks, H., Spekkink, D., and Pel, W. (2012). *BIM QuickScan: Benchmark of BIM Performance in the Netherlands*. Proceedings of CIB W78 2012: 29th International Conference on Application of IT in the AEC Industry, CIB, Rotterdam, Netherlands.

Gough, T. (2015). *BIM Certification Is a Recipe for Confusion*. Available at: www.bimplus.co.uk/people/bim-certification-recipe-confusion. [Accessed 26 March 2017].

H.M. Government (2014). *Digital Built Britain: Level 3 Building Information Modelling Strategic Plan*. Available at: www.digital-builtbritain.comDigitalBuiltBritainLevel3BuildingInformationModellingStrategicPlan.pdf. [Accessed 30 May 2015].

Kam, C., Senaratna, D., Xiao, Y., and McKinney, B. (2013). *The VDC Scorecard: Evaluation of AEC Projects and Industry Trends*. Maharashtra, India: CIFE.

NIBS (2007). *National Building Information Modeling Standard (NBIMS): Version 1.0-Part 1: Overview, Principles, and Methodology*. Washington, DC: National Institute of Building Sciences.

Ofori, G., and Gang, G. (2001). ISO 9000 certification of Singapore construction enterprises: Its costs and benefits and its role in the development of the industry. *Engineering, Construction and Architectural Management*, 2, pp. 145–157.

Succar, B. (2009). Building information modelling maturity matrix. *Handbook of Research on Building Information Modelling and Construction Informatics: Concepts and Technologies*, IGI Global, pp. 65–103.

Succar, B., Sher, W., and Williams, A. (2013). An integrated approach to BIM competency assessment, acquisition and application. *Automation in Construction*, 35, pp. 174–189.

Wu, C. Xu, B., Mao, C. and Li, X. (2017). Overview of BIM maturity measurement tools. *Journal of Information Technology in Construction (ITcon)*, 22 (3), pp. 34–64.

ANSWERS TO SET QUESTIONS AND CASE STUDIES

Here follow the answers to the various questions and case studies posed throughout the chapters of the book.

Chapter 1

Question 1

Define the following terms:

1a) Quality Policy
1b) Quality Objectives
1c) Quality Assurance
1d) Quality Control
1e) Quality Audit
1f) Quality Plan

Answer

1a) Quality Policy: policy includes the quality objectives, level of quality required by the organisation and the allocated roles for organisational employees in carrying out policy and ensuring quality. Further it shall be supported and implemented by senior organisational management.

1b) Quality Objectives: objectives are a critical component of the quality policy and for example may include establishing the competences required of staff and any associated training, in line with quality policy.

1c) Quality Assurance: Kerzner (2001) defined quality assurance as a "collective term for the formal activities and managerial processes that are planned and undertaken in an attempt to ensure that products and services are delivered at the required quality level".

1d) Quality Control: Quality control can be defined as "a collective term for activities and techniques, within the process, that are intended to create specific quality characteristics". In other words, it will assure that the organisation's quality objectives are being met, by using certain techniques such as continually monitoring processes and statistical process control (Kerzner 2001).

1e) Quality Audit: It is "an independent evaluation performed by qualified personnel that ensures that the project is conforming to the project's quality requirements and is following the established quality producers and policies" (Kerzner 2001).

1f) Quality Plan: This is a specific quality plan written for a specific project. The plan should contain the key elements/activities of the project and explain in sufficient detail exactly how they are to be delivered and assured.

Question 2

The concept of Total Quality Management has been simplified to four aspects (Haigh and Morris 2001). Identify the four aspects of TQM.

Answer

1   TQM is a total system of quality improvements with decision making based on facts rather than feeling.
2   TQM is not only about the quality of the specific product or service but it is also about everything an organisation does internally to achieve continuous performance improvement.
3   TQM assumes that quality is the outcome of all activities that take place within an organisation, in which all functions and all employees have to participate in the improvement process. In other words an organisation requires both Quality Systems and a Quality Culture.
4   TQM is a way of managing an organisation so that every job and every process is carried out right first time every time. The key to achieving sustainable quality improvement is through the adoption of TQM principles.

Question 3 – Case Study

You have been asked to act as an external consultant for 'Monaghan and Monaghan Developments' (M&M Developments). M&M Developments are considering the implementation of a formal TQM system with a view to obtaining externally verified ISO accreditations. M&M Developments consider accreditation to be a necessity in order to be placed on tender lists and continuously improve their operations.

   As an external consultant, you are requested to prepare and deliver a presentation to the senior partners of 'Monaghan and Monaghan Developments'. The topic of the presentation is 'the benefits of TQM and the associated implementation process'. Prepare notes to facilitate this presentation.

Answer

TQM can be advocated as a solution for organisations that are under-performing due to their use of traditional organisation structures and management practices whilst operating within a dynamic environment. The implementation of a TQM philosophy can facilitate performance in such organisations.

The advantages of applying a TQM approach are:

- the production of a higher quality product/service through the systematic consideration of client's requirements;
- a reduction in the overall process/time and costs via the minimisation of potential causes of errors and corrective actions;
- increased efficiency and effectiveness of all personnel with activities focused on customer satisfaction; and
- improvement in information flow between all participants through team building and proactive management strategies.

TQM can assist in making effective use of all organisational resources, by developing a culture of continuous improvement. This empowers senior management to maximise their value-added activities and minimise efforts/organisational energy expended on non-value-adding activities.

TQM enables companies to fully identify the extent of their operational activities and focus them on customer satisfaction. Part of this service focus is the provision of a significant reduction in costs through the elimination of poor quality in the overall process. This empowers companies to attain a truly sustainable competitive advantage. TQM provides a holistic framework for the operational activities of enterprises. If a firm can overcome the problematic issues of implementation, then a sustained competitive advantage is the reward to be gained.

The TQM implementation process is outlined within *Figure 1.11*.

TABLE A.1 Case study ratio answers

Ratio	2010	2009
Liquidity		
$\dfrac{Current\ Assets}{Current\ Liquidity}$	$\dfrac{1,464,000}{1,073,000} = 1.36$	$\dfrac{1,344,000}{973,000} = 1.38$
$\dfrac{Quick\ Assets}{Current\ Liabilities}$	$\dfrac{663,000}{1,073,000} = 0.62$	$\dfrac{540,000}{973,000} = 0.55$
$\dfrac{Debtors}{Sales} + 365$ (days)	$\dfrac{663,000}{2,230,000} \times 365 = \dfrac{108.51}{\text{days}}$	$\dfrac{540,000}{2,500,000} \times 365 = \dfrac{78.84}{\text{days}}$
$\dfrac{Sales}{Stocks}$	$\dfrac{2,230,000}{90,000} = \dfrac{24.78}{\text{Times}}$	$\dfrac{2,500,000}{182,000} = \dfrac{13.74}{\text{Times}}$
Profitability		
$\dfrac{P}{CE} \times 100$	$\dfrac{-20,000}{2,230,000} \times 100 = 2.03\%$	$\dfrac{119,000}{107,000} \times 100 = 10.75\%$
$\dfrac{P}{Sales} \times 100$	$\dfrac{-20,000}{2,230,000} \times 100 = 0.03\%$	$\dfrac{119,000}{2,500,000} \times 100 = 4.76\%$
$\dfrac{Sales}{CE}$	$\dfrac{2,230,000}{987,000} = \dfrac{2.26}{\text{Times}}$	$\dfrac{2,500,000}{1,107,000} = \dfrac{2.26}{\text{Times}}$

Note: * profit taken as after tax

Chapter 2

Case study question

From the data provided, Ms Smith, the Managing Director of Smith's PLC, has asked you to explain the current financial situation of her company. She has requested that you use *liquidity* and *performance* ratios as the basis of your analysis.

Answer

Liquidity comments

- The long-term liquidity ratio of 2:1 (the theoretical requirement) is not being achieved; for both time periods, 1 = approx 1.3:1.
- The acid test shows a worse scenario, the 1:1 is a useful measure. However, the figure for 2010 is only 0.62:1 – i.e. only 62 pence for every £1 of demand.
- This ratio demonstrates that the company had a problem with debt collection in 2009. But the situation has become more critical in 2010. It is taking on average of over 3 months to recover debts.
- The through-put of stocks has improved, thus less capital is tied in the company.

Performance comments

- The profit to capital employed has deteriorated from 2009–2010. The company has moved from 10.75 percent to a loss of 2 pence on every £1 employed.
- The profit generated by every £1 of sales was only 4.76 pence in 2009. However, in 2010 it has reduced to 0.59, thus the company is losing 0.89 on every £1 employed
- The capital employed is being worked at 2.26 times for both periods.

Chapter 3

Question 1

The Construction Industry can be divided into five broad sectors where quality assurance is applicable; identify these sectors.

Answer

- Client in the production of the project brief.
- Designer in the design and specification process.
- Manufacturers in the supply of materials, products and components.
- Contractors (and subcontractors) in construction, supervision and management processes.
- User in the utilisation of the new structure.

Question 2

By implementing a certified quality management system, a construction organisation can demonstrate that it has considered and deployed suitable strategies for addressing eight key quality management principles. What are these eight key quality management principles?

Answer

- **Customer-focused organisation** – organisations depend on their customers and therefore should understand current and future customer needs, meet customer requirements and strive to exceed customer expectations. This will provide a valuable assurance to potential customers.
- **Leadership** – leaders should establish an organisational unity of purpose, direction and the appropriate internal environment for the organisation, directed at customer satisfaction. They create an environment in which people can become fully involved in achieving the organisation's objectives. One of which has to be satisfying its clients.
- **Involvement of people** – people at all levels are the essence of an organisation and their full involvement enables their abilities to be used for the organisation's benefit, and hence meet customer expectations.
- **Process approach** – a desired result is achieved more efficiently when related resources and activities are managed as a process matched with customer demands.
- **System approach to management** – identifying, understanding and managing a system of interrelated processes for a given objective contributes to the effectiveness and efficiency of the organisation. Thus a methodological approach is adopted in the delivery of a quality product and or service.
- **Continual improvement** – continual improvement is a permanent objective of the organisation.
- **Factual approach to decision making** – effective decisions are based on the logical and intuitive analysis of data and information, based upon stakeholder feedback.
- **Mutually beneficial supplier relationships** – mutually beneficial relationships between the organisation and its suppliers enhance the ability of both organisations to create value. This value may then be passed on to its customers.

Question 3 – Case Study

The senior management of a construction company has been considering the deployment of an externally certified quality assurance system, as a means of potentially being included on more client tender lists.

The managing director appreciates the value of *quality* as a potential competitive advantage, and is a quality advocate within the organisation.

If the implementation of an externally certified quality assurance system is to be successful, why is it essential to have senior management support for the deployment process, and what are the likely outcomes if this support is not forthcoming?

Answer

If senior management support is not provided, the individual or team charged with the implementation of the quality system is likely to experience problems with regard to:

- inadequate authority to carry the initiative forward and bring it to a successful conclusion;
- insufficient funding for the project, and thus not being able to adequately resource the project;
- insufficient time allocation for the project thus people do not have the time to contribute; and
- resistance to:

  - information and documentation gathering;
  - implementation during the project; and
  - maintaining the system.

Successful deployment is dependent upon the strong commitment and involvement of senior management, overtly demonstrated through policies and support.

If companies are to avoid problems relating to resource issues, senior management need to provide the necessary resources; the two most important resource issues are those of adequate funding for the project and the allowance of time for people to participate in the project. Participation is necessary when the quality facilitator/project leader is gathering information for writing of the appropriate documentation. Participation of staff is also vital during the data collection and implementation phase of the project. This fact has to be recognised by and allowed for by senior management.

It should be noted that funding and time allocation are not mutually exclusive. A lack of funds can mean that money is not available to release staff when participation is requested. Also the issues of authority and overcoming resistance to change are not mutually exclusive.

Staff should be delegated sufficient authority to complete their delegated tasks. Senior management should, therefore, make sure that managers are not asked to perform tasks for which they have not been given the necessary authority and or training.

Chapter 4

Question 1

The European Foundation for Quality Management (EFQM) has stated that the functions of their Excellence Model may be split into four components. Identify these four component parts.

Answer

- As a framework which organisations can use to help them develop their vision and goals for the future, in a tangible and measurable way;
- As a framework which organisations can use to help them identify and understand the systemic nature of their business, the key linkages and cause and effect relationships;
- As the basis for the European Quality Award, a process which allows Europe to recognise its most successful organisations and promote them as role models of excellence from which others can learn; and
- As a diagnostic tool for assessing the current health of the organisation.

Question 2

The advantages of utilising EFQM.E.M's self-assessment methodology have been noted by Castka et al. (2003). Identify the advantages of EFQM.E.M's self-assessment methodology.

Answer

Benefits of using EFQM/self-assessment:

- Providing the opportunity to take a broader view on how the measured activity is impacting on the various business operations.
- Measuring performance of processes, enablers and their relationship with organisational results.
- Self-assessment conducted both internally and externally to the organisation.
- Providing an opportunity to benchmark and compare like for like or;
- Measurement for providing improvement rather than for hard quality control; and
- Self-assessment is also an important communication and planning tool:
  - The results of self-assessment provide a growing common language through which organisations, or parts of organisations, can compare their performances.
  - The outputs of self-assessment are used for strategic management and action planning, or as a basis for an improvement project.
  - New business values: leadership, people, process management, the use of information within the organisation and the way customer relationships are managed.

Question 3

The EFQM.E.M is based and supported by specific concepts which are referred to as the "Fundamental Concepts of Excellence". Identify the noted Fundamental Concepts of Excellence:

Answer

- Results orientation
- Customer focus

- Leadership and constancy of purpose
- Management by process and facts
- People development and involvement
- Continuous learning, improvement and innovation
- Partnership development
- Corporate social responsibility

Question 4 – Case Study

Answer

Key bullet points for presentation to be consulted upon completion of your own list:

Key benefits of the EFQM model have also been recognised

- It covers all areas of the organisation – offering a holistic approach, which has been absent from many other management approaches that have been used previously.
- It provides for a process of self-assessment against a non-prescriptive but detailed set of criteria, yet is flexible as to when and how this is undertaken. The approach can be adapted to suit the requirements of the user, the size of the organisational unit and the extent to which resource can be committed.
- The assessment process is based on factual evidence, but the process can be defined at a time and pace to suit the individual organisation. A self-assessment can be completed in as little as a day or with extensive evidence being collected which can take several weeks.
- It offers a means by which other initiatives such as BS EN ISO 9001:2000 can be held and knitted together in an integrated way.
- It offers a way in which a common focus can provide a new way of working that could be embedded into the organisation.
- It provides a balanced set of results indicators, not just financial, that focus on the need of the customer, the people in the organisation, the local community and other elements of society, the regulatory bodies and the funding providers.
- As the Model is used widely across Europe and has been extensively tested in a range of sectors, private, public and voluntary, it offers benchmarking opportunities with others within and outside the sector, providing a common language to share good practice and develop both individual and organisational learning.
- It provides a framework through which the kernel of the organisation's issues can be exposed, investigated and improved – continually.

The Model also engages organisations in an analysis of stakeholders, and particularly supports the recognition of the needs and expectations of customers and customer groups. The EFQM defines customers as the "final arbiter of the product and service quality, and customer loyalty". It suggests that retention and market share gain are best optimised through a clear focus on customer needs. In other words it encourages institutions to have a clear focus on the student experience.

The Model therefore offers a strong stakeholder-focused approach – which is at the heart of everything. Unless firms are driven by a way of working that looks inside at what is being done and how it is being done for all key stakeholders, then it is unlikely that continual improvement which meets or exceeds stakeholders' expectations could be achieved and sustained.

Question 5

Identify and list the nine key enabler and results criteria of the EFQM.E.M.

Answer

1 Leadership
2 People
3 Strategy
4 Partnerships and resources
5 Processes, products and services
6 People results
7 Customer results
8 Society results
9 Business results

Chapter 5

Question 1

Define the terms 'intrinsic' and 'extrinsic' motivation.

Answer

Intrinsic motivation

This is derived by fulfilling your own needs, and is therefore achieved from work itself. A considerable weight of behavioural scientific research has been devoted to the pursuit of this concept. The importance of providing feedback to employees must be understood and undertaken by managers.

Extrinsic motivation

This is deriving satisfaction of needs using work as a means to an end. It is sometimes termed the 'instrumental approach'. Work provides us with money and money enables us to 'buy' satisfaction to a certain extent. So pay is the main motivator in this line of thought.

Question 2

State the advantages of adopting a post-modernist philosophy.

Answer

The application of a post-modernist approach to managing companies can provide the following advantages:

- Organisations are more flexible and therefore better able to cope with the demands of a changing and challenging work environment.
- Teamwork and participation are attained at all levels of the company.
- Organisational culture is highly motivated and proactive.
- Corporate innovation is enhanced.
- Product/service quality is improved.
- There is a greater market awareness and thus enhanced stakeholder satisfaction.

The identified characteristics of the post-modernistic company are essential for an organisation to be able to operate both efficiently and effectively in a dynamic and turbulent operational environment.

Question 3

Define the terms *single loop*, *double loop* and *triple loop*, with regard to organisational learning.

Answer

It can be stated that in 'single-loop learning', people's decisions are based solely upon observations, while in 'double-loop learning', decisions are based on both observation and thinking.

In 'triple-loop learning', a reflection phase is incorporated to support or improve the thinking phase and hence to improve the decision-making process.

"Thus both double and triple-loop learning can be considered as generative learning, while single-loop learning can be considered an adaptive learning" (Dahlgaard 2004).

Question 4 Case Study: deploying the MFA model

Identify the organisational benefits to be obtained from deploying the MFAM.

Answer

Resulting Benefits

- A clear understanding of how to deliver value to clients and hence gain a sustainable competitive advantage via operations.

- Enabling the mission and vision statements to be accomplished by building on the strengths of the company.
- Ability to gauge what the organisation is achieving in relation to its planned performance targets.
- Clarity and unity of purpose so that the organisation's personnel can excel and continuously improve.
- Interrelated activities are systematically managed with a holistic approach to decision making.
- People development and involvement. Shared values and a culture of trust, thus encouraging empowerment in line with post-modernist company.

The Management Functional Assessment Model (MFAM) provides a focal point for those managers seeking to be proactive in the management of change processes. The MFAM also combines six functions that are usually considered in a disparate fashion, if at all. Identify the six noted functions.
  Six functions

- Setting and implementing strategic plans;
- Setting and implementing operational plans;
- Giving due consideration to organisational size, when selecting and engaging in self-assessment linked to an improvement model;
- Linking the various functions of management in an effective and efficient way;
- Obtaining feedback for stakeholders on organisational performance, with a view to the enhancement of service and product provision; and
- Building on the concept of triple-loop learning.

Chapter 6

Question 1

Identify the benefits associated with the deployment of an OHS management system.

Answer

Benefits can include:

- Improved prevention of occupational injury and disease – a safer and healthier workplace.
- The provision of a framework for identifying hazards and managing the resultant risks.
- A reduction in the loss of working days due to accidents and injury.
- A reduction in the incidences of employee compensation claims.
- The development of a reviewable approach for meeting legislative requirements, duties of care and due diligence.
- A reduction in insurance premiums.

- Improved morale and productivity brought about by employee inclusivity with developing and running.
- Enhanced working methods that facilitate improvement in production and productivity rates.
- Enhanced reputation of the organisation with a visible and tangible commitment to continuous improvement and inclusive, consultative management mechanism.
- Reduced staff turnover and thereby reduced 'replacement costs'.
- Improved ability to attract skilled personnel.
- Improved commercial potential – inclusion on tender lists is increased as the potential for meeting the pre-qualification requirements of significant clients is enhanced.

Question 2

The Institution of Occupational Safety and Health (IOSH) outlines a process for the development within an organisation of an occupational health and safety management system. Six typical inputs are identified within this process. Identify these six typical inputs.

Answer

The six typical inputs identified are:

1   Any information relating to hazard identification and risk assessment.
2   Review of OSH performance, including incidents and accidents.
3   Identification and review of existing OSH management arrangements or processes.
4   Competence and training requirements.
5   Workforce involvement.
6   OSH legal and other standards and best practice within sector, e.g. a compliance register.

Refer to *Figure 6.2* for further information.

Question 3

You have been asked to deliver a brief presentation at the next senior management meeting of your department. The presentation concerns an upcoming H&S audit. The title of the presentation is: 'An Outline of the Key Inspection Components of the Upcoming Health and Safety Self-Assessment Audit'.

Answer

Key components of the self-assessment audit should include:

- Safety policy
- Organisation for safety control

- Organisation for safety competence
- Organisation of workforce safety involvement
- Organisation of safety communication
- Safety planning and implementation
- Safety performance measurement
- Auditing and reviewing safety.

Refer to *Table 6.4* for further information and detail regarding the subcomponents of the audit.

Chapter 7

Question 1

What are '5D' and '6D' BIM?

Answer

The focus of 5D BIM is commercial management '5D BIM' caters to estimating, cost management and procurement. It includes 'time-cost-value' analysis techniques such as Earned Value Management. Work is also under way to integrate BIM applications with enterprise resource planning (ERP) systems at the business level of the organisation to inform sales, purchasing and logistics functions.

6D BIM concerns sustainability. It allows information such as energy use, resource efficiency and other aspects of sustainability to be better analysed, managed and understood. The BIM model can accommodate information such as embodied carbon, including that created by the process of construction.

Question 2

The key requirements for a BIM-enabled project have been identified as including?

Answer

- Employer's Information Requirements (EIR), which sets out a client's requirements for the delivery of information by its project supply chain.
- BIM Execution Plan (BEP) or Project Execution Plan (PEP), which demonstrates how the EIR will be delivered and which can contain a Master Information Delivery Plan (MIDP) to indicate when project information is to be produced, by whom and how.
- The Project Information Model (PIM) consisting of all the documentation, non-graphical and graphical information that defines the delivered project and which (for the purpose of managing, maintaining and operating the asset) is eventually superseded by the Asset Information Model (AIM).
- A Common Data Environment (CDE) that, amongst other things, will contain all of the above.

Question 3

What is COBie ?

Answer

COBie is a data schema presented in the form of a spreadsheet which serves as a standardised index of information about new and existing assets throughout their life cycle. BS 1192–4 Collaborative production of information. Part 4: Fulfilling employers' information exchange requirements using COBie: This represents a revision of BS 1192:2007 to encompass the handling of information using COBie.

Question 4

The UK Government Cabinet Office has proposed three "new models of procurement" that would best correspond to "high levels of supply chain integration, innovation and good working relationships. Identify these three 'new models of procurement".

Answer

- *Cost-led procurement*: The most conventional of the three 'new models' in which an 'integrated framework supply team' is selected from up to three competing bids, based on affordability and quality criteria.
- *Two-stage open book*: A first stage, in which contractor-consultant teams compete on the basis of a development fee and qualitative elements is followed by a second, where the successful team openly develop the project proposal to the client's cost benchmark.
- *Integrated Project Insurance*: Following competition based upon qualitative criteria and a 'fee declaration', an integrated project team is selected to develop an acceptable design solution and a single joint-names project insurance policy is executed to cover all risks associated with delivery of the project.

Chapter 8

Question 1

Project suppliers are required to answer BIM policy and capability compliance questions contained in PAS91.

Identify four BIM policy and capability compliance supplier questions that project suppliers are required to answer.

Answer

The four supplier questions are:

1   Do you have the capability of working with a project using a "Common Data Environment" as described in PAS 1192:2:2013?

2 Do you have documented policy, systems and procedures to achieve "Level 2 BIM" maturity as defined in the government's BIM Strategy?

3 Do you have the capability of developing and delivering or working to (depending upon the role(s) that this PQQ covers) a BIM Execution Plan (BEP) as described in PAS 1192:2:2013?

4 Do you have arrangements for training employees in BIM-related skills and do you assess their capabilities?

Question 2

What are the exemption criteria of PAS 91:2013 that relate to BIM, policy and capability?

Answer

The exemption criteria of PAS 91:2013 that relate to BIM, policy and capability are:

- that there is a certificate of compliance with BS PAS 1192:2:2013 and
- that the certificate of compliance is from a third party organisation with a related UKAS accreditation, or equivalent.

Question 3

Name three organisations that offer some form of recognised BIM certification that purports to satisfy these PAS exemption criteria.

Answer

- BRE Global (part of the Building Research Establishment),
- The British Standards Institute (BSI), and
- LRQA (part of the Lloyd's Register Group).

INDEX